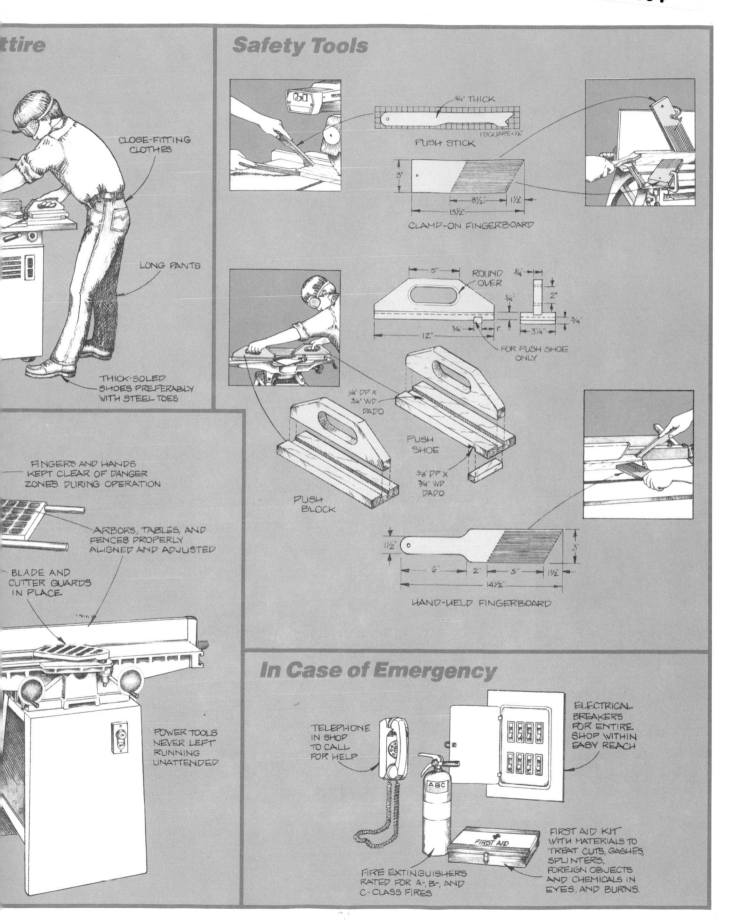

ttire

CLOSE-FITTING CLOTHES

LONG PANTS

THICK-SOLED SHOES PREFERABLY WITH STEEL TOES

FINGERS AND HANDS KEPT CLEAR OF DANGER ZONES DURING OPERATION

ARBORS, TABLES, AND FENCES PROPERLY ALIGNED AND ADJUSTED

BLADE AND CUTTER GUARDS IN PLACE

POWER TOOLS NEVER LEFT RUNNING UNATTENDED

Safety Tools

¾" THICK

PUSH STICK

1 SQUARE = ½"

3"

8½"   1½"

13½"

CLAMP-ON FINGERBOARD

5"

ROUND OVER

¾"

¾"

2"

12"

¾"

r

3¼"

¾"

FOR PUSH SHOE ONLY

¼" DP X ¾" WD DADO

PUSH SHOE

⅜" DP X ¾" WD DADO

PUSH BLOCK

1½"

3"

6"   2"   5"   1½"

14½"

HAND-HELD FINGERBOARD

In Case of Emergency

ELECTRICAL BREAKERS FOR ENTIRE SHOP WITHIN EASY REACH

TELEPHONE IN SHOP TO CALL FOR HELP

ABC

FIRST AID

FIRST AID KIT WITH MATERIALS TO TREAT CUTS, GASHES, SPLINTERS, FOREIGN OBJECTS AND CHEMICALS IN EYES, AND BURNS.

FIRE EXTINGUISHERS RATED FOR A-, B-, AND C-CLASS FIRES

·BUILD·IT·BETTER·YOURSELF·
WOODWORKING PROJECTS

Outdoor Structures

Collected and Written
by Nick Engler

Rodale Press
Emmaus, Pennsylvania

Printed in the United States of America

If you have any questions or comments concerning this book, please write:
Rodale Press
Book Reader Service
33 East Minor Street
Emmaus, PA l8098

Series Editor: Jeff Day
Managing Editor/Author: Nick Engler
Editor: Roger Yepsen
Copy Editor: Mary Green
Graphic Designer: Linda Watts
Graphic Artists: Mary Jane Favorite
                  Chris Walendzak
Photography: Karen Callahan
Cover Photography: Mitch Mandel
Cover Photograph Stylist: Janet C. Vera
Proofreader: Hue Park
Typesetting by Computer Typography, Huber Heights, Ohio
Interior Illustrations by O'Neil & Associates, Dayton, Ohio
Endpaper Illustrations by Mary Jane Favorite
Produced by Bookworks, Inc., West Milton, Ohio

Library of Congress Cataloging-in-Publication Data

Engler, Nick.
    Outdoor structures/collected and written by Nick
Engler.
        p.   cm.—(Build-it-better-yourself
    woodworking projects)
    ISBN 0–87857–845–5 hardcover
    1. Garden structures—Design and construction—
    Amateurs' manuals.
    I. Title.   II. Series: Engler, Nick. Build-it-better-
yourself woodworking projects.
TH4961.E54   1990
684—dc20   89–24236
                                                    CIP

2 4 6 8 10 9 7 5 3 1   hardcover

Contents

Outdoor Building: Plans and Materials

If every person's home is indeed a castle, then the surrounding yard is a kingdom — sovereign land on which you may build anything you please (provided it satisfies the local building codes). By adding a few well-chosen outdoor structures, you can improve the appearance of your property, increase its value, create recreational and storage areas, and expand your warm-weather living space. In short, you can tailor the property to your personal needs and tastes, making it uniquely yours.

You must tailor each outdoor project, too — to fit your family's needs, your budget, your experience, the available space and materials, the surrounding landscape and architecture, and a dozen other criteria. You probably won't build any of the outdoor structures exactly as they're presented in the following chapters, since each was designed for someone else's yard. Instead, you'll adapt them to your own personal vision, changing the size and design as needed. To do this, you must carefully *plan* your structure and understand the *materials* that are available to you.

Planning

Before you begin any outdoor building project, you should do some research and, if necessary, a little paperwork. Construction requirements vary from place to place, and you may have to adapt your plans to suit local codes, neighborhood committees, geography, or climate.

Checking codes and other regulations — Before you make any plans, consult the local building and zoning authority. Depending on where you live, this may be part of the county, city, or township government. Consult the listing of "Government Offices" in your phone book. (You can usually find this in the blue pages.) Call or visit the office and briefly explain what you want to do. They will tell you what codes and restrictions apply, and whether or not you need a building permit.

You may also want to check with any local neighborhood committees, particularly if you live in a historic district. Many neighborhoods set their own architectural standards, either to preserve local historical traditions or to beautify the area. These don't have the force of law, as do building codes, but it may be worth your while to comply with them anyway. A building project that meets or exceeds local standards adds more to the property value than one that doesn't.

Don't presume that codes and standards only apply to large projects. Smaller ones may also require some research. For instance, a fence only a few sections long will have to meet local building codes. Even birdhouses have standards — the U.S. Department of the Interior recommends different sizes and configurations, depending on the species you want to attract.

Drawing plans — Once you have researched local codes and any other applicable standards, give some thought to your own requirements — where you should place the structure, how big it should be, the arrangement of doors and windows (or other components), and the overall architectural design. Then draw up a set of plans.

These don't have to be formal architectural drawings, like the plans in this book. But you should at least make some sketches and note the dimensions. Plans help you visualize what a project will look like so you can adjust the proportions and the design before you build. They also help you to think through a project — go through the construction steps in your mind — so you can estimate the amount of lumber and other materials that you will need.

Selecting Materials

You can build most outdoor structures from ordinary construction-grade lumber and hardware. With only a few exceptions, these materials are available at a lumberyard or building supply center.

Lumber species and types — Most construction-grade wood is either Douglas fir, southern pine, or western red cedar. In some locations, you may also find ash, spruce, white pine, redwood, larch, cypress, gum, and hemlock. All of these woods are good choices for outdoor building projects, depending on how you use them.

Most woods used *aboveground* must be painted, stained, or placed under a roof so they will stand up to the weather. A few species — redwood, cypress, and cedar — are naturally resistant to rot and decay, and can be exposed to the elements. You can also buy decay-resistant *pressure-treated* lumber. This is pine or fir that has been impregnated with a toxin (usually chromated copper arsenate) to prevent the growth of bacteria that rot the wood. Pressure-treated boards are rated according to how much of the toxin they retain — the more toxin, the better the board will hold up to the weather. The most common ratings are LP-2 (aboveground use only) and LP-22 (suitable for ground contact). Neither needs to be painted, although you can

paint pressure-treated lumber. If you do, allow at least two months after construction for the wood to season.

When selecting lumber, consider where you will use it in a project and how it will be covered or finished. Also consider climate and location — how wet is the area where you intend to build? Then select a wood that's appropriate to the condition.

Condition	Lumber
Aboveground, painted and/or under roof	All types
Aboveground, unpainted and/or exposed	Redwood, cypress, cedar, LP-2
Extremely wet location and/or ground contact	Redwood, cypress, cedar, LP-22
Exposed to salt water or spray	Cedar, LP-88

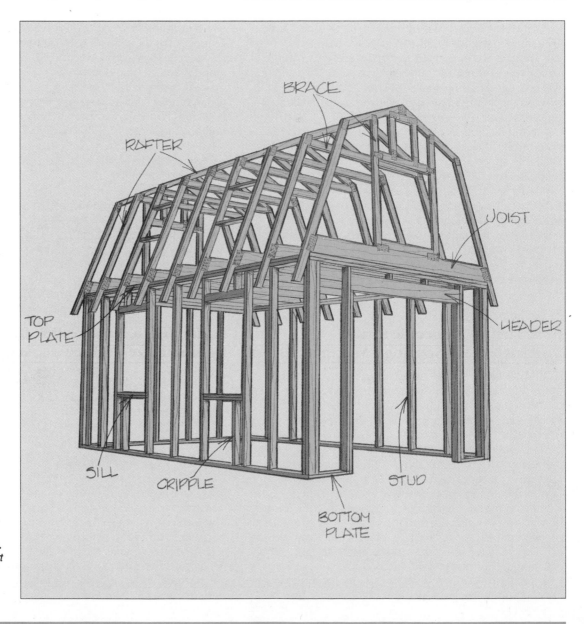

*For the most part, the structural members of a building make up the **frame**. These parts support most of the weight of the building and keep it rigid.*

Lumber sizes — After choosing the type of lumber, decide the sizes. Construction lumber is sawed and surfaced to standard *nominal board sizes* — 1 x 6, 2 x 4, 4 x 4, etc. The nominal board size needed for a specific application depends on how much weight a board must support, how it's spaced in relation to other boards, and how long a span it has to bridge.

Local building codes will dictate lumber sizes, and you should check these before purchasing materials. However, most codes follow these simple guidelines: For most vertical and angled structural members (studs, cripples, braces, and rafters in trusses), use 2 x 4s. Don't space these more than 24″ apart, on center. You may also use 2 x 4 lumber for some horizontal structural members — sills, plates, and purlins. Size the other members (beams, headers, joists, planks, and rafters that aren't parts of trusses) according to their span and spacing.

Plywood — Plywood is used in outdoor building for roof sheathing and siding, and occasionally for flooring.

No matter what you use it for, you should always specify exterior plywood — ACX, BCX, or CDX. As with interior plywood, the first two letters describe the quality of the front and back surfaces, with "A" at the top end of the scale and "D" at the bottom. The "X" signifies that the plies have been bonded with a waterproof glue and won't delaminate when exposed to the weather.

The thickness of the plywood depends on what you use it for. Often, the plywood not only covers the frame of a building, it also *braces* it. In other words, it supplies much of the structural strength. The more strength you need, the thicker the material should be. For example, you can side or sub-side small buildings with ⅜″-thick plywood. However, larger structures require ⅝″. Floors and lofts always need ¾″, to better support the weight. You can sheath most roofs with ½″ plywood, but you may want to use ⅝″ or even ¾″ in areas with heavy snowfalls. You may also want to use ¾″-thick material to prevent the points of roofing nails from coming through the sheathing.

Beam and Header Size for a Specific Span

Nominal Size	Maximum Span*
4 x 4 (or two 2 x 4s)	4′
4 x 6 (or two 2 x 6s)	6′
4 x 8 (or two 2 x 8s)	8′
4 x 10 (or two 2 x 10s)	10′
4 x 12 (or two 2 x 12s)	12′
6 x 10 (or three 2 x 10s)	12′
6 x 12 (or three 2 x 12s)	14′

With boards placed on edge

Plank Size for a Specific Span

Nominal Size	Maximum Span*
1 x 4	12″
1 x 6	16″
2 x 2	42″
2 x 4	42″
2 x 6	42″

With boards placed flat

Joist Size for a Specific Span

Nominal Size	Spacing	Maximum Span*
2 x 6 (minimum)	12″ o.c.	8′
	16″ o.c.	7′
	24″ o.c.	5′
2 x 8	12″ o.c.	10′
	16″ o.c.	9′
	24″ o.c.	7′
2 x 10	12″ o.c.	13′
	16″ o.c.	12′
	24″ o.c.	10′
2 x 12	12″ o.c.	16′
	16″ o.c.	15′
	24″ o.c.	14′

With boards placed on edge

Rafter Size for a Specific Span

Nominal Size	Slope	Maximum Span*
2 x 4	15°	6′
	30°	7′
	45°	9′
2 x 6	15°	9′
	30°	10′
	45°	13′
2 x 8	15°	11′
	30°	13′
	45°	16′
2 x 10	15°	14′
	30°	16′
	45°	20′

With boards placed on edge

Use	Conditions	Type	Thickness
Flooring	All floors, sub-floors, or lofts	BCX	¾″
Sub-siding	Small building (one story, 8′ x 8′ or less)	CDX	⅜″ or more
	Large building	CDX	⅝″ or more
Siding	Small building	ACX or BCX	⅜″ or more
	Large building	ACX or BCX	⅝″ or more
Roofing	Most climates	CDX	½″
	Areas of heavy snowfall	CDX	⅝″ or ¾″
	When appearance of underside of roof is important	BCX (B side down)	¾″

You also may use exterior-grade particleboard or chipboard. If you do, remember that neither is as strong as plywood, and won't brace a building as well as the same thickness of plywood. You should use material that is ⅛″ to ¼″ thicker than shown in the chart above. Also remember that particleboard is less moisture resistant than exterior plywood, even though it may be rated for outside use. Despite the rating, you shouldn't use it in wet climates.

Hardware — If the nails, screws, and other hardware on an outdoor project will be protected from the weather, use ordinary iron or steel. For example, the frame of a storage shed can be assembled with common, untreated nails, since you will cover it with roofing and siding.

However, if hardware will be left exposed to the weather, purchase items that won't rust or corrode. This is true even if you intend to cover the fasteners with paint. Rainwater will soak underneath paint, rust the nails, and stain the project. In addition, the corrosion will eat away at the nails, and the structure will begin to fall apart. To avoid this, look for rustproof or weatherproof hardware. You have several choices:

■ *Galvanized* nails and hardware are your least expensive option. There are two types — plated and hot-dipped. Of the two, hot-dipped hardware will better weather the elements.

■ *Brass* screws and fasteners won't corrode when exposed to *fresh* water. However, they may disintegrate around salt water. There are other drawbacks. Brass is a soft metal; screw heads are easily stripped, and hinges are easily bent. Brass hardware is also expensive. Avoid brass-plated items; these are not weatherproof.

■ *Bronze* hardware is extremely hard, and will stand up to both fresh water and salt water. But it's very expensive and not readily available. You may have to special-order most items, or purchase them at a marine supply store. Don't purchase bronze-plated hardware; it's not weatherproof.

■ *Stainless steel* is hard and stands up to both fresh and salt water. It's slightly less expensive than bronze.

■ *Zinc- or cadmium-plated* hardware is rust resistant, but not really weatherproof. Use it only if your climate is dry.

Roofing — The projects in this book are roofed with asphalt tab shingles laid over roofing paper. There are several important advantages to these materials: They're good-looking, readily available, inexpensive, relatively simple to install, long-lived, and easy to adapt to a variety of roof shapes and surfaces. However, there are other types of roofing that you may wish to consider:

■ *Roofing felt* or "roll roofing" is installed in long, 36″-wide strips. It's perhaps the easiest and quickest of the roofing materials to install, but it doesn't last very long and it isn't much to look at.

■ *Sheet metal* roofs will last longer than almost any other sort of roofing. They are expensive, but spare you the need to sheathe the roof with plywood, and this offsets some of the expense. You can apply the metal sheets easily to simple gabled or shed roofs, but they are hard to adapt to out-of-the-ordinary shapes, such as gazebo roofs.

■ *Cedar shakes* are more aesthetic than the other types of shingles mentioned here, but they are also more difficult and time-consuming to install. They're quite expensive, but will outlast other roofing materials if properly installed.

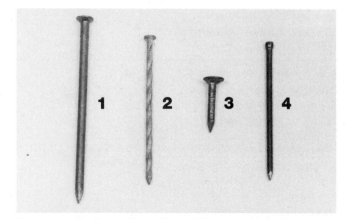

*Four types of nails are frequently used in outdoor projects. Assemble frames and structural members with **common nails** (1). These have a long shank and a medium-size head, and are suitable for general assembly. Attach flooring, decking, and siding with **spiral nails** (2), also called patio or decking nails. The square, twisted shank ensures that they won't easily work themselves loose. Attach mending plates and roofing materials with **roofing nails** (3). These have a short shank and a large head to hold thin materials (such as shingles and sheet metal) to the wood. Finally, attach window frames, doorjambs, and most trim with **finishing nails** (4). These have a long shank and a tiny head that can barely be seen after you drive them into the wood. You can purchase untreated common, spiral, and finishing nails. All four types of nails are available galvanized.*

Multipurpose Frame Building

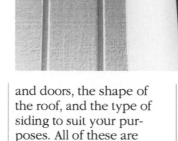

What sorts of out-buildings do you need on your property? A large storage barn for a garden tractor? A shed to dry wood? A finishing room, away from the dust in your shop? A detached garage? A playhouse for the children? A screened-in summerhouse for bug-free picnics? A rustic vacation cabin? An open shelter from which you can watch birds and wildlife?

All of these structures can be built in a similar fashion. They fall into the same general size range — about the proportions of a single room in your home. You won't live in them, so they won't be subjected to the same rigorous building codes that apply to a house. None of them requires heating, electricity, or plumbing, although you can add utilities if you need them. Nor do they require insulation or interior finish work, although you can add that too, if you're so inclined. In short, any of them can be constructed as a basic, no-frills frame building.

The building shown is a wood-drying shed with a storage loft. But you can adapt the plan to make any one of the structures mentioned — and many more. You only have to change the overall size, the locations of windows and doors, the shape of the roof, and the type of siding to suit your purposes. All of these are *minor* changes. No matter how they may alter the appearance of the building, the construction procedure will remain the same.

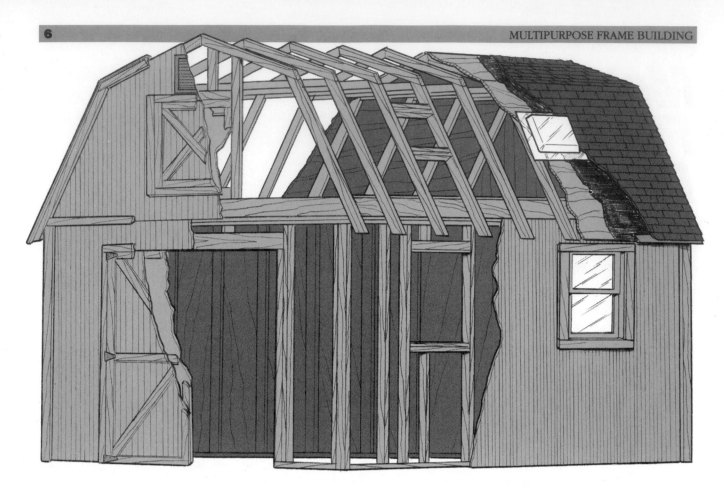

CUTAWAY VIEW

Materials List

Amount	Material	Use
9	2 x 8 x 10′	Joists
2	2 x 8 x 8′	Door header
6	2 x 4 x 16′	Sill plate, top plate, cap plate
5	2 x 4 x 10′	Sill plate, top plate, cap plate
80	2 x 4 x 8′	Studs, corner posts, rafters, braces, window headers and sills, cripples, purlins
18	⅝″ x 4′ x 8′ Exterior plywood siding	Siding
5	¾″ x 4′ x 8′ BC plywood	Loft floor
12	½″ x 4′ x 8′ CDX plywood	Roof sheathing
6	¼″ x 4′ x 8′ BCX plywood	Barn and loft doors
15	1 x 12 x 10′ #2 pine	Trim, spacers, blocks

HARDWARE

15 lbs.	16d Common nails	Assemble frame
3 lbs.	10d Common nails	Assemble heads, sills, plates
1 lb.	10d Finishing nails (galvanized)	Install windows, door stops
10 lbs.	8d Spiral nails (galvanized)	Attach siding, sheathing, trim, jambs, loft floor

(Continued)

Materials List — *Continued*

Amount	Material	Use
1 lb.	4d Common nails (galvanized)	Assemble doors
5 lbs.	1¼" Roofing nails	Attach mending plates, rafter ties
10 lbs.	¾" Roofing nails	Attach roofing materials
122	Mending plates	Assemble trusses
36	Rafter ties	Attach trusses to cap plate
8	10' Drip-edge strips	Roofing
1 roll	Roofing paper	Roofing
366 sq. ft. (11 bundles)	Asphalt tab shingles	Roofing
2	24" x 24" Skylights	Illuminate loft
2	30" x 42" Single-pane windows	Illuminate interior
2	12" x 18" Vents	Ventilate loft
3 pr.	6" Strap hinges	Mount doors
1 pr.	6" T-hinges	Mount loft door
24	#10 x 2½" Flathead wood screws	Mount hinges to barn
32	#10 x 1¾" Stove bolts, flat washers, and nuts	Mount hinges to doors
1	Hasp and mounting screws	Secure door

Note: *This list was figured for the storage barn* **as shown in the drawings.** *It does not include a floor frame or flooring. Depending on the type of foundation you choose, and the size and configuration of the structure you build, you may have to adjust the amounts of materials.*

1 Determine the purpose, location, size, and configuration of the building.

Decide what sort of a building you want to build. A storage barn? Playhouse? Picnic shelter? A combination? Do some research and find out what local codes and restrictions apply to this kind of outbuilding, and whether or not you need a building permit.

Look around your yard and decide where to put the building. This, too, may be regulated by local codes. Most planning and building agencies require you to place a *permanent* building at least 10 feet in from your property line. You may also have to erect the new building a certain number of feet away from existing structures, wells, septic fields, and utility lines. Sketch a plan of the building, showing the location of windows, doors, vents, skylights, stairs, and ramps.

Also sketch a framing plan, similar to those shown — framing plans are invaluable aids. They come in handy not just for cutting and construction, but also for *estimating*. Using the plans, compile a list of the lumber and hardware needed.

TRY THIS! If the plan calls for studs that are *slightly* shorter than 8', consider buying *precut* or *basement* studs. Precut studs are 92⅝" long, and basement studs are 84" long. This will reduce the amount of waste — and save money — when you cut the framing parts. To build the storage barn shown, for example, you could substitute basement studs for 61 of the 80 required 2 x 4s.

2 Build a foundation and floor.

Decide what sort of foundation to put under your building. The barn shown is built on a shallow pier-and-beam foundation, but you can also use deep piers or a concrete pad. The type of foundation will affect the list of materials. For example, if you build a barn on shallow piers as shown, you'll also need:

- 2   2 x 10 x 16′
- 9   2 x 10 x 10′
- 16   Sets of cross bridges
- 20   5/4 x 6 x 16′ Decking
- 12   8″ x 8″ x 16″ Solid concrete blocks

These materials will build the foundation, floor frame,

and floor shown. A concrete pad will require less lumber, but more hardware and other materials.

When you decide on a foundation, add the necessary materials to your list and purchase them all. Build the foundation and, if necessary, the floor frame and floor. (Refer to "Step-by-Step: Building a Pier-and-Beam Foundation," page 22 or "Step-by-Step: Pouring a Concrete Pad," page 66.)

3 Build the roof trusses.

It might seem that the next logical step after building the foundation and floor would be building the walls. However, that is not necessarily the case. The floor provides a flat, open surface on which you can build *both* the wall frames and the roof trusses — if you build them in the proper order. The wall frames are large, weak assemblies that have little strength until you join and brace them — you should erect them as you build them. But when you put them up, this open area disappears. The trusses, by comparison, are somewhat smaller and a great deal stronger. They can be easily set aside and stored until you're ready to use them. So it's usually best to build the trusses first, then the wall frames.

Cut the parts needed for the trusses, lay them out on the floor, and join them with mending plates. If you've built a wooden floor, use it to simplify this task and

ensure that all the trusses are precisely the same. Build the first truss, then nail 2 x 4 scraps to the floor, all around and inside the truss. Arrange the scraps in pairs, each pair butting against the inside and outside edge of a truss member. Flank each member with at least four scraps — two at each end. This will form a huge truss-building jig. (See Figure 1.) Set the first truss aside, and lay the parts for the second in the jig. Join the parts with mending plates, turn the truss over, and nail mending plates to the other side. Repeat for each truss.

Build the end trusses last. Remove the scraps that flank the interior parts of the truss (braces), but keep those for the exterior parts (rafters and joists). Put all the parts in place and join them as before. (See Figure 2.) This allows you to frame in doors and vents on the end gable while ensuring that the end trusses have the same exact shape as those in the middle.

1/With some well-placed 2 x 4 scraps, you can turn a wooden floor into a huge truss-building jig. Nail the scraps directly to the floor, then remove them after you've built all the trusses.

2/When building the end trusses, remove all but the exterior blocks from the jig. This will allow you to frame in vents, windows, and doors as needed, while retaining the same exact **outside** shape as the middle trusses.

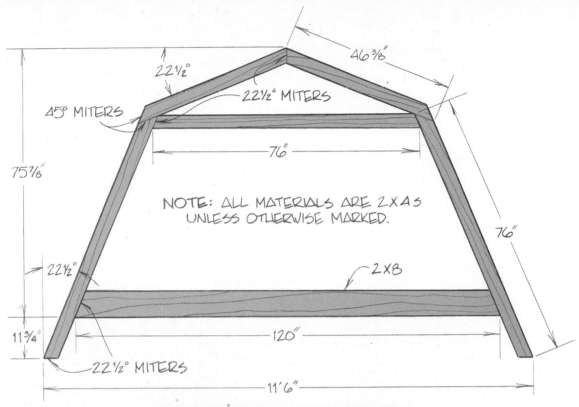

STORAGE BARN TRUSS (MAKE 7)

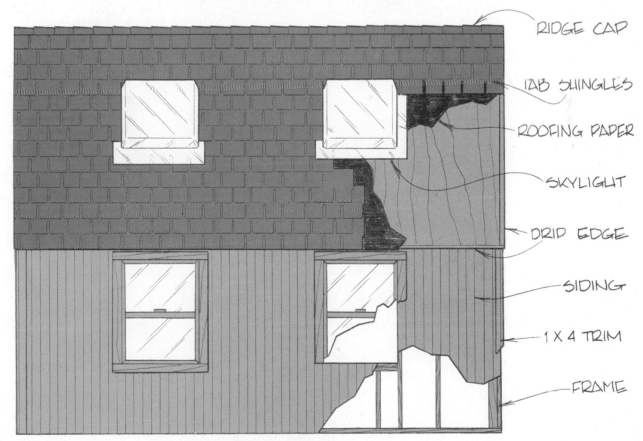

SOUTH ELEVATION

4 Build and erect the wall frames.

Remove all the scraps from the floor. Build the wall frames, and erect them one at a time. Start with one of the long frames — cut the parts, nail it together, and stand it up in place. Tack it or bolt it to the floor or foundation, but don't attach it permanently until you're sure you won't need to shift the position of the frame slightly. Have a helper hold the frame plumb, then brace it upright.

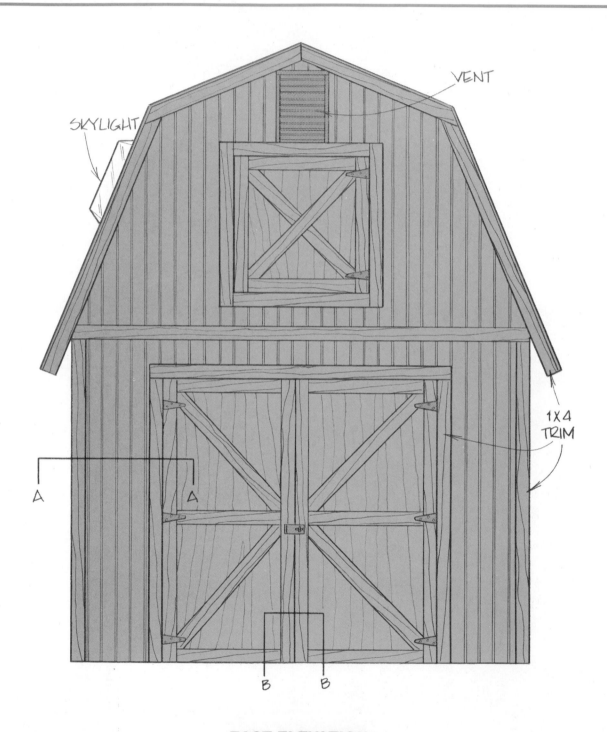

EAST ELEVATION

Next, build an adjoining frame. Stand this in place and nail it to the first frame to form a corner. Attach it to the floor and brace it as you did before. (See Figures 3 through 6.)

At this point, you may want to skip ahead and build the fourth and last frame you plan to erect: If you build the third frame and put it in place, there may no longer be enough room on the floor to build the fourth one. So build this last frame, set it aside, then build the third. Plan your work so the fourth frame is the smallest and lightest of all the wall frames. Erect the third frame, then the fourth.

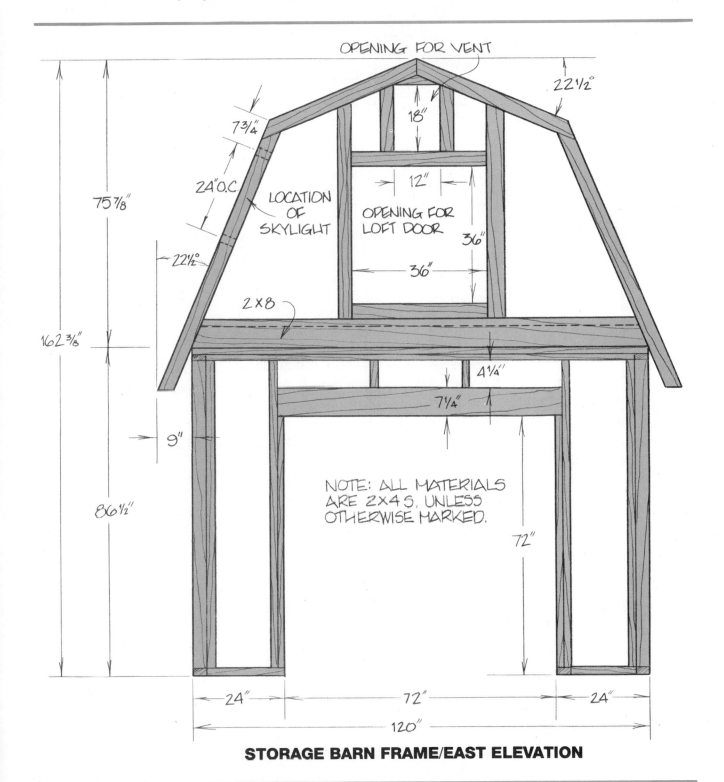

STORAGE BARN FRAME/EAST ELEVATION

3/Build and erect each frame in four steps. First, cut the parts, lay them on the floor or foundation, and nail them together.

4/With a helper, stand the frame up in place. If the frame is slightly out of alignment when it's upright, use a mallet to tap it into position.

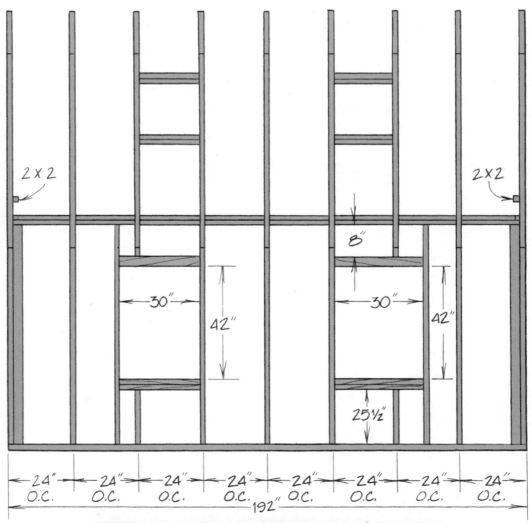

2 X 2

2 X 2

8"

30"   30"

42"   42"

25½"

24" O.C. 24" O.C. 24" O.C. 24" O.C. 24" O.C. 24" O.C. 24" O.C. 24" O.C.

192"

STORAGE BARN FRAME/SOUTH ELEVATION

5/Join the corners of the frame to the adjoining frames, driving 16d nails through the corner posts. Then nail or bolt the bottom plate to the floor.

6/Finally, brace the frame so it's plumb and square. Temporarily nail a 2 x 4 diagonally from the corner across several studs. Repeat this process for each successive wall frame.

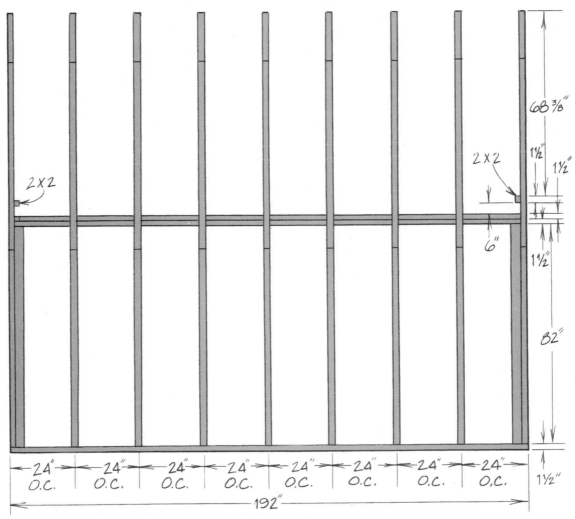

STORAGE BARN FRAME/NORTH ELEVATION

5 ***Tie the wall frames together.*** When all the frames are properly positioned and braced, attach them to the floor permanently. Then nail cap plates to the top plates — the cap plate members must span the joints between the wall frames. (See Figure 7.) This will tie the frames together.

7/Tie the wall frames together with cap plates. This prevents the weight of the roof from pushing the corners apart.

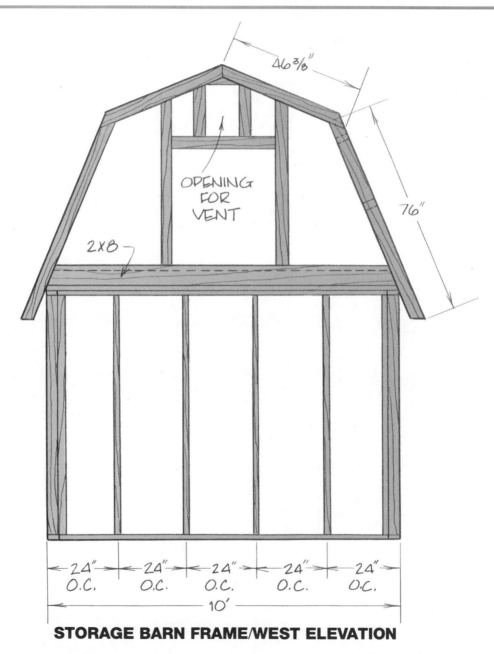

STORAGE BARN FRAME/WEST ELEVATION

6 Erect the roof trusses.

Place the roof trusses on top of the cap plate, attach them with rafter ties, and brace them upright. If your structure incorporates fairly large trusses, you'll find it's easier to put the trusses in place upside down, then flip them upright. (See Figures 8 through 11.) Nail long scraps of wood to the rafters, spanning the trusses. This ties the trusses together and keeps them properly spaced until you attach the sheathing.

8/To erect a large truss, first put it in place between the walls upside down. The top edge of the rafter should rest on the cap plate.

9/Rotate the truss upward, as if it were hinged to the cap plate. Use a long pole to help push the truss into position. Then attach the truss to the cap plate and brace it upright.

10/You won't be able to flip the last few trusses into place — there won't be room. Instead, slide the trusses up over the wall frames, as shown.

11/Raise the truss until it's upright. Once again, use a large pole to push it into position. Note the long scrap that ties the trusses together.

7 **Install the loft floor (optional).** If your plan calls for a loft, install the loft floor now. Nail a 2 x 2 ledger to the two end trusses, where shown in the *Storage Barn Frame/South Elevation.* The top edges of the ledgers must be flush with the top edges of the joists.

Using 8d spiral nails, attach the plywood or the flooring to the joists. (See Figure 12.) Be sure to leave an open area through which you can enter the loft from the ground floor. You may also want to build and install a ladder to get up to and down from the loft easily. (See Figure 13.)

*12/Install the loft floor **before** you sheath the roof. Be careful to position the seams between the plywood sheets over the joists.*

13/Make a simple ladder to reach the loft — this will save you time as you sheath the roof. If you wish, the ladder can remain as a permanent part of the building.

8 **Install the roof sheathing.** If you plan to add skylights, install 2 x 4s between the trusses to frame the skylight openings. Attach plywood sheathing to the trusses with 8d spiral nails, removing the temporary ties and braces as you work. (Refer to "Step-by-Step: Installing Tab Shingles," page 101.) *Don't* cover the skylight openings. (See Figure 14.)

14/Cover the trusses with sheathing. If you plan to install skylights, leave these areas uncovered.

9
Install the siding.   Cover the wall frames
with siding. If you're using board siding or sheet
siding, simply attach the pieces with 8d spiral nails.
(See Figure 15.) For lap siding, install sub-siding
material, then nail the lap boards over it.

*15/Install sheet
siding in much the
same way you
installed roof
sheathing — just
nail it to the frame.
Leave areas open
for the doors, win-
dows, and vents.*

10
Attach the jambs, trim, and facing.
Using 8d nails, attach the jambs inside the door
opening. These jambs should cover the frame and the
edges of the siding. Then attach 1 x 4 trim around the
edges of the doors, the edges of the roof, the corners,
and across the ends of the building. The jambs and trim
are shown in the *East Elevation, South Elevation,* and
Section A. Also attach 1 x 4 facing to the bottom ends of
the rafters. Bevel the outside edge of the facing at 22½°
to match the slope of the roof. (See Figure 16.)

*16/Before you
attach the roof trim
or facing, install
siding on the ends
of the eaves, as
shown. Although
this siding won't be
seen, it serves as a
spacer for the roof-
edge trim.*

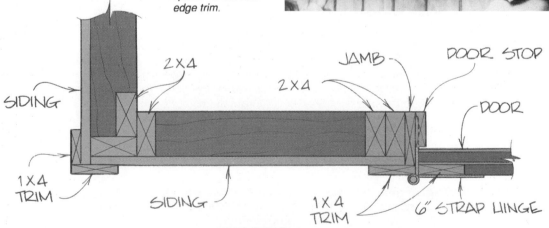

SECTION A

11 **Install the roofing materials.** Cover the sheathing with roofing paper, using ¾″ roofing nails to hold the paper down. Install drip-edge strips around the perimeter of the roof, then cover both the paper and the drip edge with tab shingles. (Refer to "Step-by-Step: Installing Tab Shingles," page 101.)

Install the skylights at the same time that you install the shingles. When the rows of shingles reach the bottom edge of the skylight openings, attach the skylights to the rafters and 2 x 4s with 1¼″ roofing nails. (See Figure 17.) Continue to apply the shingles, lapping them over the flanges of the skylight. (See Figure 18.)

*17/As the shingles reach a skylight opening, nail the skylight in place. The **bottom** flange of the skylight must lap **over** the shingles.*

*18/Install the remaining shingles around the skylight. When all the shingles are in place, the **side** and **top** flanges of the skylight should be **under** the shingles.*

12 **Install the windows and doors.** Place the windows in their openings, square the units, and attach them to the frame with 10d finishing nails following the manufacturer's instructions. Do the same for the vents.

Carefully measure the door openings. Cut the spacers and sheets needed to make core doors for both the loft and the ground floor openings. Using 4d nails, assemble the sheets and spacers. Arrange the spacers as shown in the *Loft Door Spacer Layout* and the *Door Spacer Layout*. Then attach the trim as shown in the *Loft Door Detail* and the *Door Detail*. When attaching the trim to the lower doors, set the outside vertical trim piece about ¾″ in from the edge of one door; on the other door, attach the trim piece so it laps the edge of the first door, as shown in *Section B*. Hang the doors on strap hinges, as shown in the *Door Hanging Detail*. Hang the loft door on T-hinges.

13 **Build a stair (optional).** If necessary, build a stair unit (refer to "Step-by-Step: Building Stairs," page 48) or a ramp in front of the doors. Ramps are made in much the same manner as stairs, but the run should be much longer than the rise for an easier slope. And don't cut notches for the treads in the stringers. Leave the top edges straight and nail planks across them.

14 **Paint the building.** Prime the raw, exposed surfaces of the siding, facing, trim, doors, and windows. Then paint the building. You can also treat the wood with a weatherproof stain.

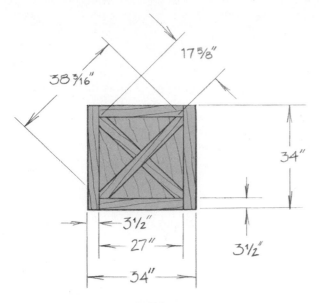

LOFT DOOR DETAIL

Dimensions: 38 3/16", 17 5/8", 34", 3 1/2", 27", 3 1/2", 34"

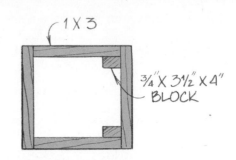

1 X 3

3/4" X 3 1/2" X 4" BLOCK

LOFT DOOR SPACER LAYOUT

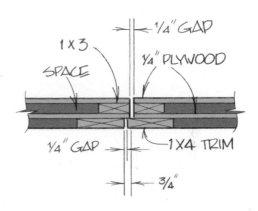

1/4" GAP
1/4" PLYWOOD
1 X 3
SPACE
1/4" GAP
1 X 4 TRIM
3/4"

SECTION B

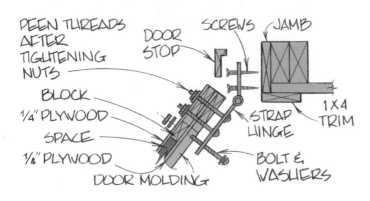

PEEN THREADS AFTER TIGHTENING NUTS

DOOR STOP

SCREWS

JAMB

BLOCK

1/4" PLYWOOD

SPACE

1/4" PLYWOOD

DOOR MOLDING

STRAP HINGE

BOLT & WASHERS

1 X 4 TRIM

DOOR HANGING DETAIL

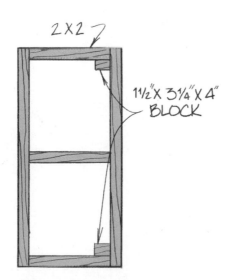

2 X 2

1 1/2" X 3 1/4" X 4" BLOCK

DOOR SPACER LAYOUT

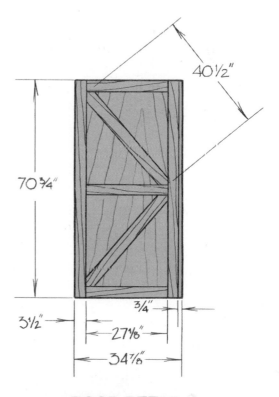

40 1/2"

70 3/4"

3 1/2"

3/4"

27 1/8"

34 7/8"

DOOR DETAIL

Variations

As mentioned in the beginning of this chapter, you can modify this building plan to serve many purposes. As shown, the building has a gambrel roof with solid walls. This makes a good shed, storage barn, or garage.

For a playhouse or cabin, you'll probably want to build a structure with a gable roof. To do this, build gable trusses as shown in the *Gable Truss Layout*.

To make a picnic shelter, you'll have to change the walls. Build wall frames with interior bracing, as

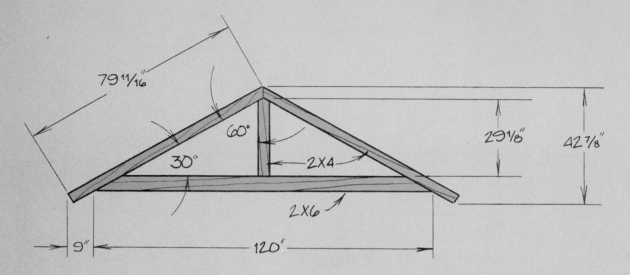

GABLE TRUSS LAYOUT

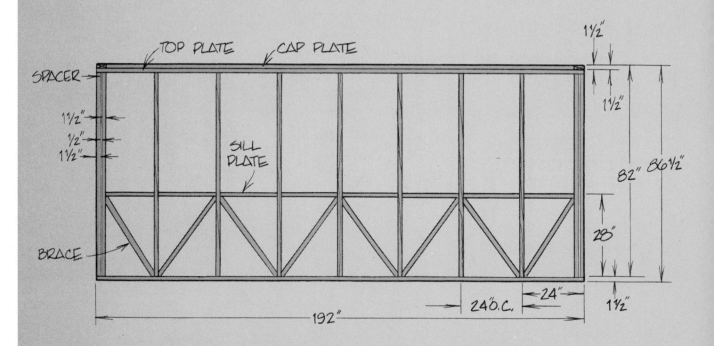

PICNIC SHELTER FRAME/SOUTH ELEVATION

shown in the *Picnic Shelter Frame*. Since you want both the inside and the outside of the building to look good, install siding on both sides of each wall, but only part way up the wall. Leave the upper two-thirds of the wall frame open, as shown in the *Picnic*

Shelter Wall/Section View. Install jambs and spacers around the openings, tack screen over the jambs, and hide the edges of the screen with molding, as shown in the *Picnic Shelter Corner/Section View* and *Picnic Shelter Construction Detail*.

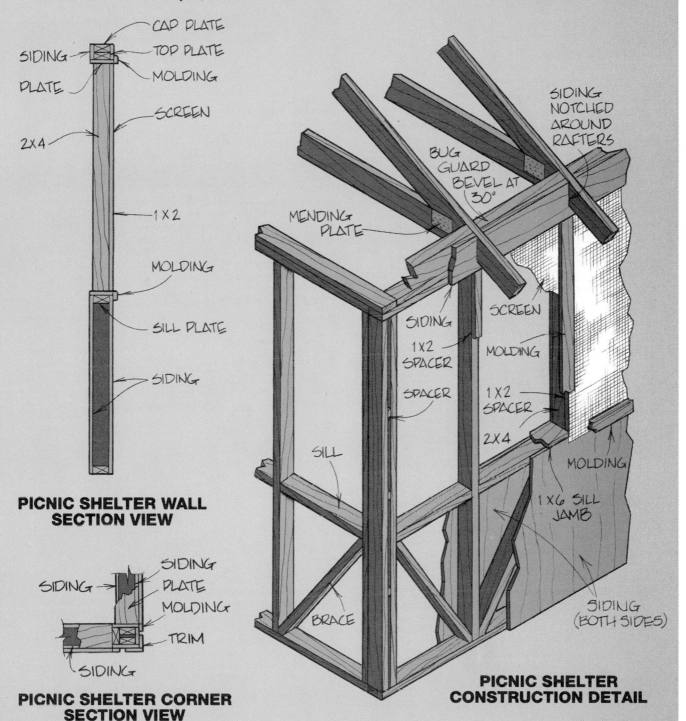

**PICNIC SHELTER WALL
SECTION VIEW**

**PICNIC SHELTER CORNER
SECTION VIEW**

**PICNIC SHELTER
CONSTRUCTION DETAIL**

Step-by-Step: Building a Pier-and-Beam Foundation

Pier-and-beam foundations consist of wooden beams or joists stretched across concrete piers. Unlike other foundations, these piers don't have to be sunk deeply into the ground, nor do the beams have to be attached to the piers. This creates a unique advantage: In many locations, a small building on shallow piers qualifies as a nonpermanent structure. This exempts it from property taxes. Also, since it can be moved, the building may not have to satisfy many of the building codes that apply to permanent structures. You can place a barn or deck closer to the edge of your property than you might otherwise, and build it from cheaper materials. (It will still have to meet some local standards, however, and you should check with your local building and zoning agency to find out what these are.)

Use *solid* 8″ x 8″ x 16″ concrete blocks for the piers. These are not available at most lumberyards or building centers. You may either have to special-order them or buy them from a masonry supplier or fabricator. Look in the yellow pages of your telephone book under "Concrete Products."

You can also install a permanent pier-and-beam foundation, using *deep piers* instead of shallow ones. Cast these deep piers from cement right in the ground.

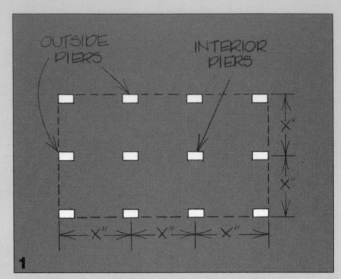

1

Carefully plan the location of the piers according to the size of the beams. Also consider the weight of the building and the condition of the soil. For light buildings or clay soil, space the piers further apart. For heavy buildings or sandy soil, space them closer together. Remember to place piers to support the **interior** of the frame, where necessary. If you're in doubt as to how far apart to space the piers, it's always better to have too many than too few.

4

Nail the four outside beams together on the corner piers. Check that the beams are square to each other by measuring diagonally from corner to corner in the same manner that you checked the foundation layout. When the beams are square, brace each corner with 2 x 4s. If necessary, tap the piers in place so the outside of each pier is flush with the outer faces of the beams that rest on it. Then fill the shallow holes around each pier with dirt and tamp it down. This will keep the corner piers from moving.

5

Put the remaining piers in place under the beams. First raise one side of the frame and brace it 12″–16″ above the piers. Then stretch strings across the top and outside edges of two corner piers. Install the piers that lie between these corners so they just touch the strings. Remove the strings and braces, lower the frame back on the piers, and repeat for the other sides. Also install any interior piers. As you work, periodically check that the frame remains level when resting flat on the piers. When you're finished, tap the outside piers in place so they are flush with the outer faces of the beams. Fill in around the piers and tamp the earth down.

Lay out the location of the corner piers, using a measuring tape and ordinary concrete blocks. Place the blocks where you want the corners, then measure diagonally from the outside corner of one block to the outside corner of another. Repeat for the other two blocks. The two diagonal measurements should be the same. If not, shift one or more of the blocks, until the measurements are the same.

Install the corner piers. Dig a shallow hole for the first pier, 4"–8" deep. Tamp the earth at the bottom of the hole and put the pier in place. Dig a hole for the second pier, put it in place, and lay a beam across both piers. Check that the beam is level. If not, raise or lower the second pier by filling in the hole or digging it deeper. Repeat for the remaining corner piers.

Nail the joists in place between the beams. If needed, use joist hangers to suspend the beams. Attach X-braces (also called cross bridges) between the joists. These help to distribute the load across the frame and to make the floor more rigid. Those shown are ready-made metal braces, but you can also make your own from scrap wood.

Install the decking, flooring, or subflooring across the beams and joists with spiral nails. Don't bother to cut the planks to length as you go. You can work faster and do a neater job if you wait until you've installed all the floorboards, then cut them off flush with the face of the beams.

User-Friendly Doghouse

A doghouse ought to be more than four walls and a roof. *You* need more than these bare essentials to be comfortable, so it stands to reason that your *pet* does, too. A simple shelter will shield a dog from the wind and rain, but with just a little extra time and materials, you can also protect your pet from excessive heat and cold, and — to a certain extent — ticks and fleas.

The doghouse shown looks simple enough, but its various features help keep the house warm (or cool), properly ventilated, and dry.

Heating and cooling — The house is custom-fit to the size of the dog so that the shelter can be easily warmed by the animal's body heat. The floor rests several inches above the cold, damp earth to keep the house warmer and drier. The walls, roof, and floor are insulated to help keep the house warm in the winter and cool in the summer.

Ventilation — The roof is hinged like a lid so that you can adjust the ventilation. In the winter, close it tight to keep the warm air in the doghouse. In the summer, open it several inches to let the hot air escape. The door is raised several inches above the floor to prevent cold drafts from blowing in over the sill.

Bedding — The raised door also helps keep the bedding in place. You can fill the bottom of the house with cedar chips or sawdust. To keep the bedding dry, the floor is slanted slightly toward the door. Any rain that happens to blow in will quickly drain through a gap between the floor and the front wall. (Dry bedding not only keeps the dog warm, but also discourages fleas and other pests.) To replace the bedding and clean the house, open the roof all the way for easy access to the inside.

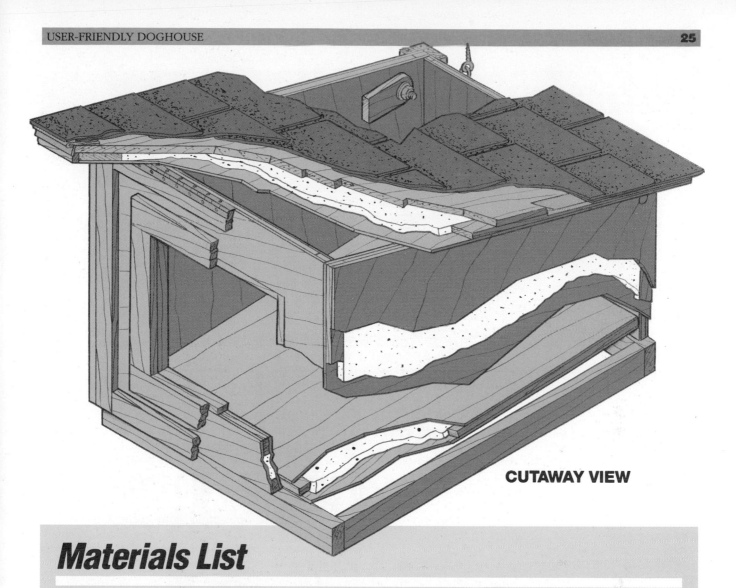

CUTAWAY VIEW

Materials List

Amount	Material	Use
4	¼" x 4' x 8' Exterior particleboard	Roof, bottom, front, back, side panels
2	1 x 12 x 8' #2 pine	Spacers, trim, facing supports
2	2 x 4 x 8' Treated LP-22	Foundation
1	¾" x 4' x 8' CDX plywood	Roof panel

HARDWARE

Amount	Material	Use
1	¾" x 4' x 8' Styrofoam insulation	Roof, bottom, front, back, side panels
20 sq. ft.	Roofing paper	Roof
33 sq. ft. (1 bundle)	Asphalt tab shingles	Roof
2	10' Drip-edge strips	Roof
6	¾" Eye screws	Secure roof
2	4" Turnbuckles	Secure roof
4	Large S-hooks	Secure roof
2	⅜" x 2½" Carriage bolts with flat washers and nuts	Roof supports
1	30" Piano hinge and mounting screws	Mount roof
¼ lb.	16d Common nails (galvanized)	Assemble foundation
½ lb.	10d Spiral nails (galvanized)	Assemble panels

(Continued)

Materials List — *Continued*

Amount	Material	Use
1 lb.	4d Common nails (galvanized)	Assemble panels
1 box	1″ Wire brads	Assemble panels
1 lb.	¾″ Roofing nails	Attach roofing materials
½ lb.	6d Spiral nails (galvanized)	Attach facing, trim
5	#12 x 1½″ Flathead wood screws	Attach trim
14′	Weather stripping and mounting tacks	Seal roof
3 tubes	Construction adhesive	Assemble panels
1 gal.	Exterior paint	Protect panels

Note: *This list of materials was figured for the project* **as shown in the drawings.** *If you change the size of the doghouse, you may have to adjust the amounts of materials.*

---

1 **Adjust the size of the doghouse to fit your dog.** All dogs are descendants of wolf ancestors, and as such they prefer close, cavelike homes. You may think you're doing your dog a favor by building a roomy doghouse, but what your pet really wants is something that's just a little bit bigger than he is. The entrance should be no bigger that it has to be, and the sill should be as far above the door as practical.

As drawn and built, this house was designed for a medium-large dog (a border collie). You will probably want to change the size somewhat for your own dog. To figure the optimum *interior* dimensions of a doghouse, measure your dog's length and width when he's *lying down* (but not curled up). Measure his height *standing up*. Add 4″ to the length and width, and 1″ to the height. To figure the width and height of the entrance, measure from the dog's chest to his head (excluding the legs). Add 2″ for small and medium dogs, 3″–4″ for large and extra-large dogs. To locate the sill, place it about half the length of the dog's legs above the floor, up to 7″–8″.

This will give the dog ample room to enter, turn around, and lie down. However, the space will be small enough to feel warm and cozy. The size and location of the entrance will help to conserve the warmth while keeping out cold air, rain, and snow.

Once you've figured these dimensions, make the necessary changes to the plan. You may also have to adjust the amounts of materials in the Materials List.

---

2 **Cut the parts for the panels.** The roof, front, back, side, and bottom panels are wood-and-Styrofoam sandwiches, to insulate the doghouse. The Styrofoam is covered on *both* sides with particleboard or plywood. This not only makes the panel rigid, but also keeps the dog from chewing the Styrofoam.

You need to cut two *identical* pieces of particleboard or plywood for each panel. The simplest way to get identical pieces is to nail two pieces together and cut them simultaneously. First lay out the parts on a sheet of ¼″ particleboard. Mark the front and back pieces about ½″ higher than you need, so you can bevel the top edges *after* the panels are assembled.

Then stack the marked-up sheet on top of an unmarked sheet. Drive a few small wire brads through each of the pieces you marked, *inside* the cut lines. Cut the sheet with a circular saw, then cut out the entrance opening in the front with a saber saw. When you remove the brads and take the sheets apart, you'll have two identical pieces for each panel. Do this for everything *except* the roof. For this one panel, lay a sheet of particleboard over a sheet of ¾″ plywood and cut them both to size.

Rip the 1 x 2, 1 x 3, and 1 x 6 spacers you need from the 1 x 12 stock. Fit the spacers to the panels and cut them to length. Where necessary, miter the ends at 15°.

TRY THIS! When cutting large sheets of plywood or particleboard, lay a sheet of ⅝″ soft fiberboard (sometimes called builder's board) on the floor of your shop or some other flat surface. This gives you much more support than sawhorses, and makes the cutting operation easier and safer. Lay the material to be cut on top of the fiberboard, and adjust the circular saw to cut through the material and just ⅛″ into the fiberboard.

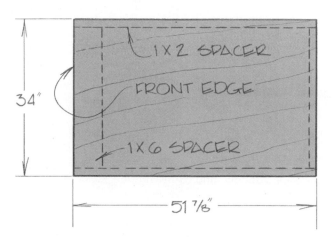

ROOF PANEL LAYOUT

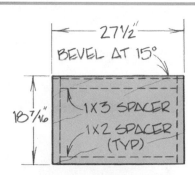

BACK PANEL LAYOUT

FLOOR PANEL LAYOUT

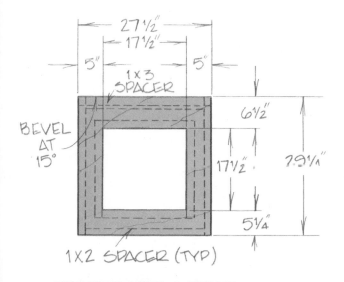

FRONT PANEL LAYOUT

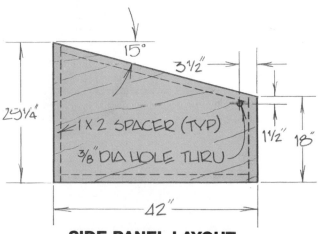

SIDE PANEL LAYOUT

3

Assemble the panels. Select a set of sheets and spacers for a single panel, and lay one sheet flat on your workbench. Apply a generous bead of construction adhesive to the face of the 1 x 2 spacers, and tack them to the sheet with wire brads, adhesive side down.

With a utility knife, cut pieces of Styrofoam to fill the space between the spacers. Insert the Styrofoam and apply another bead of adhesive to the upturned face

of the spacers. (See Figure 1.) Lay the second, identical sheet on top of the spacers and secure it to the spacers with 4d nails. Then turn the assembly over and drive a few nails in from the opposite side to secure the first sheet. Repeat for each panel, until you have made the six panels needed. (See Figure 2.)

Important: Where you have used 1 x 3 spacers, place the nails 1½"–2" back from the edge. This will give you room to cut the bevels needed on these panels.

1/Completely fill the cavity in each panel with Styrofoam — there should be no air pockets. Use construction adhesive to seal the panels so no moisture can penetrate the cavity.

2/Attach the sheets to the spacers with the smallest galvanized nails you can find. If the nails are longer than the panel is thick, drive them at an angle so they don't go completely through.

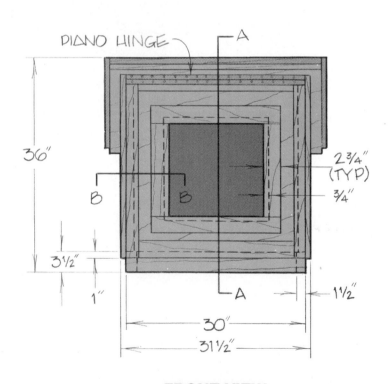

FRONT VIEW

4 **Bevel one edge of the front, back, and bottom panels.** The upper edges of the front and back must be beveled at 15°, and the back edge of the bottom at 2°, as shown in *Section A*. The 1 x 3 spacers along these edges give you some extra

stock to cut the bevels. Tilt the blade of a circular saw to the proper angle. Carefully measure the size of the panel, mark it, and cut the bevel. Be careful not to hit any nails!

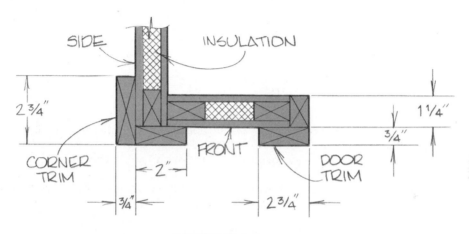

SECTION B

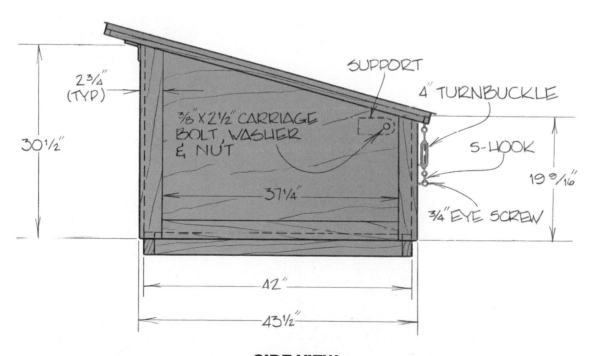

SIDE VIEW

5

Assemble the panels. Temporarily dry assemble the front, back, sides, and bottom to test the fit. Remember, the bottom should slope slightly from the back to the front, and there should be a ¼″ gap between the front and bottom panels. When you're satisfied the panels fit properly, reassemble them with construction adhesive and 10d spiral nails.

6

Mount the house on a foundation. From 2 x 4 stock, cut the parts needed to make a simple frame for the foundation. The outside dimensions of this frame should be exactly the same as the perimeter of the assembled doghouse. Assemble the frame with 16d nails and rest the house on top of it.

7

Attach the trim to the doghouse. From the remaining 1 x 12 stock, rip trim and facing. You'll need 2¾″-wide trim around the bottom, both 2″- and 2¾″-wide trim along the corners, 1¼″- and 2¾″-wide trim around the entrance, and 1¾″-wide facing around the edge of the roof.

Fit the corner trim to the doghouse first, mitering or beveling the top edges as needed. Lap the bottom ends of the trim ¾″–1″ over the foundation frame. Attach the trim to the house with 6d spiral nails, but *don't* attach it to the foundation — the foundation should remain detachable. After trimming the corners, cut and attach the bottom trim. Once again, nail these parts to the house but not to the foundation.

Trim the inside edges of the entrance opening with 1¼″-wide trim, then attach 2¾″-wide trim around the perimeter. Nail 1¾″-wide facing all around the edges of the roof. Finally, bevel one edge of a piece of 2″-wide trim at 15° and *screw* it to the front, flush with the top edge. This piece of top trim will provide a mount for the piano hinge that holds the roof in place.

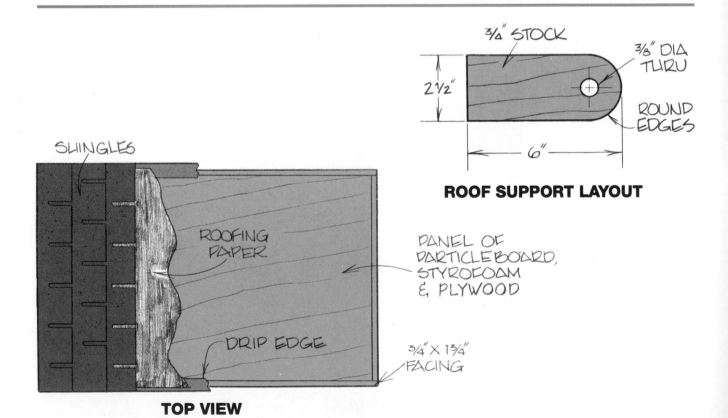

ROOF SUPPORT LAYOUT

TOP VIEW

8

Attach the roof. Cut two 2½"-wide, 6"-long supports from one-by (¾"-thick) stock. Round one end of each support, and drill a ⅜"-diameter hole through it, as shown in the *Roof Support Layout*. In addition, drill ⅜"-diameter holes through the sides of the doghouse, as shown on the *Side Panel Layout*. Attach the roof supports to the inside surface of the sides with carriage bolts, flat washers, and nuts. Tighten the nuts just enough that the supports are snug but still able to pivot on the bolts.

Tack weather stripping or an insulating gasket all around the top edge of the front, back, and sides. Tack it in only a few places — preferably, just the corners — so that you can remove it later.

Place the roof on the house so it's centered from side to side and overhangs the front by 6". Attach the roof to the doghouse by mounting the piano hinge on the underside of the roof and the top front trim. Use screws that are long enough to go through the trim and into the panels.

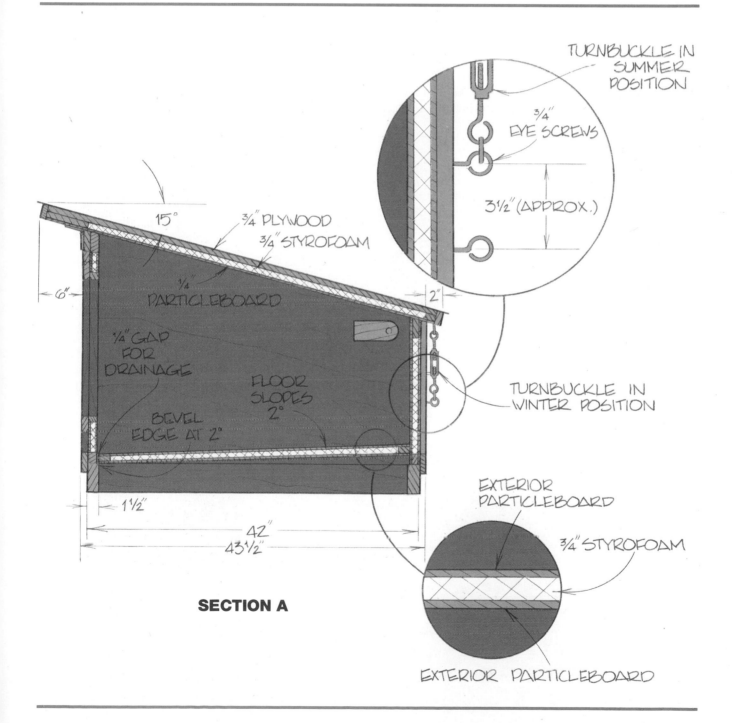

TURNBUCKLE IN SUMMER POSITION

¾" EYE SCREWS

3½" (APPROX.)

15°

¾" PLYWOOD

¾" STYROFOAM

6"

¼" PARTICLEBOARD

2"

¼" GAP FOR DRAINAGE

FLOOR SLOPES 2°

BEVEL EDGE AT 2°

TURNBUCKLE IN WINTER POSITION

1½"

42"

43½"

EXTERIOR PARTICLEBOARD

¾" STYROFOAM

SECTION A

EXTERIOR PARTICLEBOARD

9 Install the roof hold-downs.

Install two eye screws in the underside of the roof, in opposite corners and near the back edge. Using S-hooks, fasten one turnbuckle to each eye screw. Fasten an S-hook to the opposite end of each turnbuckle, but leave open the free end of the S-hook. The free end of the S-hooks attach to one of two sets of eye screws — the first set holds the roof closed for winter. The second set holds the roof partially open for summer.

Adjust both turnbuckles until they're extended about three-quarters of the way. Hold them so the lower (open) S-hooks touch the trim at each corner — this marks the position for a second set of eye screws. Install these screws, then open the roof. Turn the supports until they protrude 3"–4" above the sides, and let the roof rest on them. Hold the turnbuckles so the S-hooks meet the corner trim again, and install the last set of eye screws where the hooks touch.

10 Paint the doghouse.

Remove all the hardware from the assembled doghouse — hinge, eye screws, turnbuckles, and weather stripping. Take the roof off the house, and the house off the foundation. Paint all surfaces — inside and outside, and tops and bottoms *except* for the foundation and the top surface of the roof. Let the paint dry thoroughly, then reassemble the house. Tack the weather stripping down securely.

11 Apply the roofing materials.

Put the roof in the winter position, so it's flat against the top edges of the house. Cover the top surface of the roof with roofing paper, install drip-edge strips around the edge, and apply asphalt tab shingles. (Refer to "Step-by-Step: Installing Tab Shingles," page 101.)

12 Place the doghouse in your yard.

Choose a location for the doghouse and decide which way it will face. The roof should slope into the prevailing wind — this will help keep the house dry through rain and snow storms, ventilate it in the summer, and conserve warmth in the winter. Place the foundation on the ground and level it, so the floor will drain properly. Then place the house on the foundation.

Open or close the roof as the weather dictates. In the winter, when you want the doghouse to retain heat, turn the supports so they face down, close the roof, and hook the turnbuckles to the lower set of eye screws. Tighten the turnbuckles so the roof is tight against the gasket around the top edge. In the summer, when you want to ventilate the house, open the roof, turn the supports up, and fasten the turnbuckles to the upper set of eye screws. Tighten the turnbuckles to hold the roof securely against the supports. (See Figure 3.)

When you want to clean the doghouse — summer or winter — loosen the turnbuckles and open the roof all the way. Clamp it or brace it open so you can easily reach the inside. (See Figure 4.)

3/When turned up, the roof supports raise the roof slightly, providing better ventilation. When turned down, the roof rests against the weather stripping, sealing in warm air.

4/The roof of the doghouse tilts forward so you can easily reach the inside to clean it or change the bedding. Be careful to brace it open so it doesn't fall on you while you're working.

Nailing Fundamentals

*T*he parts of most outdoor structures are joined with nails. While nailing may seem to be a simple skill, it often requires as much careful thought as other types of joinery. Here are a few tips to help you through various nailing procedures that you may encounter.

When building any frame, or when attaching flooring, siding, or roofing to that frame, you may use several different nailing techniques. Mostly, you'll simply nail straight through one board and into the next. Whenever you can, drive the nails at a slight angle. As you drive each successive nail, alternate this angle left and right. This hooks the boards together and prevents them from coming loose.

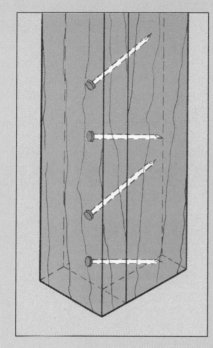

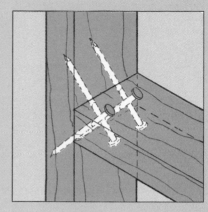

*Now and again, you may have to **toenail** two boards together — drive nails at a steep angle through the end of one board and into the face or edge of another. If possible, drive at least two nails from opposite directions.*

If the nails go completely through both boards, you have three choices: (1) Use smaller nails. If this isn't practical, (2) clinch exposed ends or (3) drive them at an angle so they don't exit the wood.

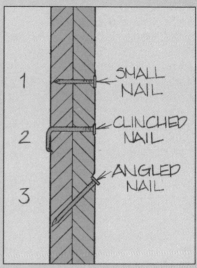

1 SMALL NAIL

2 CLINCHED NAIL

3 ANGLED NAIL

*Sometimes you must join two boards or assemblies with a third piece called a **gusset,** especially when assembling roof frames. You can make your own gussets from sheet metal or scraps of plywood, or buy them ready made. Some of the shapes available are shown here: (a) **L-brackets,** (b) **T-brackets,** (c) **mending plates,** (d) **joist hangers,** (e) **angle clips,** (f) **rafter ties,** and (g) **post anchors.** Unless otherwise instructed, attach these gussets to the wooden framing members with roofing nails.*

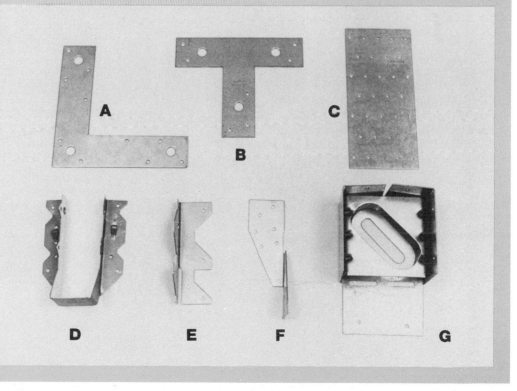

Country Deck

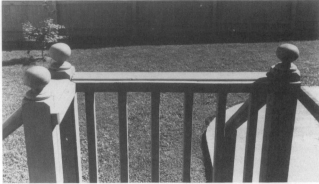

There is no easier or less expensive way to expand your living space than to add a deck to your home. A deck is just a wooden platform that extends out from the house. In warm, seasonable weather it provides a dry, level surface for picnic tables, barbecue grills, and lawn furniture. This, in effect, creates an extra room for dining, and relaxing.

The deck shown was built by Larry Callahan of West Milton, Ohio. When Larry decided to build this deck, he faced a common problem in designing it: He owns an older home, and most deck plans are quite modern, meant to be built with pressure-treated lumber. A contemporary deck would clash with his turn-of-the-century architecture. What design could be built economically from today's materials, but would still be appropriate to the older architecture?

Larry's solution was to imitate the lines of an old farmhouse porch, using readily available lumber rather than turned spindles or other traditional gingerbread. The result was an elegant "country deck" that works well with both older and newer styles of homes.

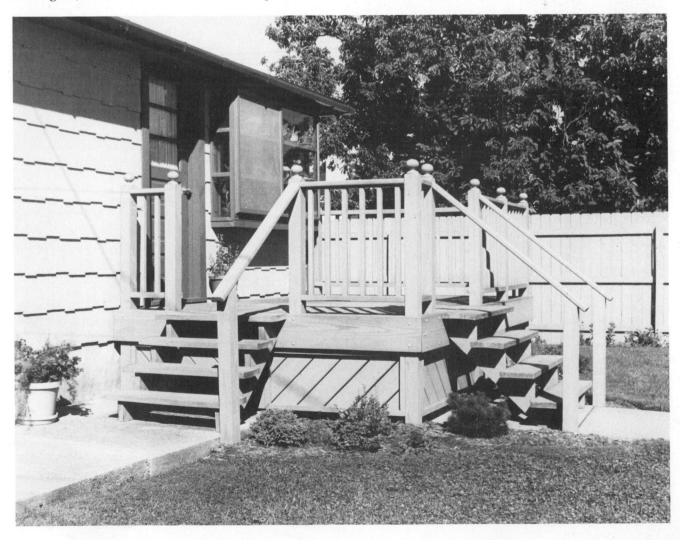

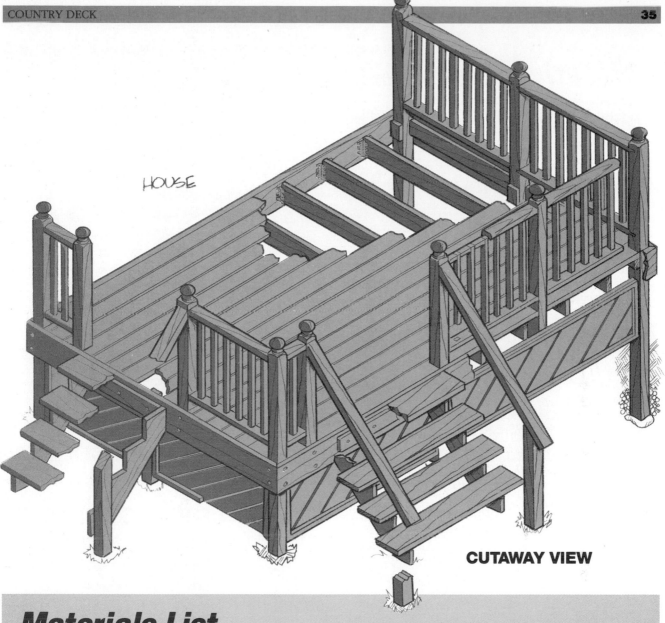

HOUSE

CUTAWAY VIEW

Materials List

Amount	Material	Use
8	4 x 4 x 8′ Treated LP-22	Supporting posts, upper newel posts, posts
2	4 x 4 x 12′ Treated LP-22	Lower newel posts
2	2 x 10 x 12′ Treated LP-2	Ledger, beams
8	2 x 10 x 8′ Treated LP-2	Beams, stair risers, treads
6	2 x 8 x 8′ Treated LP-2	Joists
1	2 x 6 x 8′ Treated LP-2	Stair ledgers
25	5/4 x 6 x 12′ Decking	Planks, skirts
7	2 x 4 x 8′ Shaped railing	Top rails
5	2 x 4 x 8′ Treated LP-2	Short ledgers, bottom rails
12	2 x 2 x 8′ Treated LP-2	Balusters
2	1 x 6 x 8′ Treated LP-2	Spacers for balusters
10	Post-end ornaments	Finials

(Continued)

Materials List — Continued

Amount	Material	Use
HARDWARE		
6–8	⅜″ x 6″ Lag screws	Attach ledger to house
24	⅜″ x 4″ Lag screws	Attach posts to frame
10	⅜″ Double-pointed lag screws	Attach finials to posts
5 lbs.	16d Common nails (galvanized)	Assemble frame, attach skirt
1 lb.	16d Spiral nails (galvanized)	Attach stair treads, hand rails
1 lb.	10d Spiral nails (galvanized)	Attach short ledgers and stair ledgers
1 lb.	10d Finishing nails (galvanized)	Attach top and bottom rails to posts
10 lbs.	8d Spiral nails (galvanized)	Attach decking
1 lb.	6d Common nails (galvanized)	Attach spacers to rails, assemble skirt
6–8	#14 x 3½″ Flathead wood screws (brass)	Attach removable skirt panel
12	Joist hangers	Attach joists to ledgers and frame
1 lb.	1¼″ Roofing nails	Mount joist hangers
8	10″ Anchor bolts	Anchor stair risers
4	Concrete blocks	Anchor stair risers

Note: *This list was figured for the project **as shown in the drawings.** If you change the size and configuration of the deck, you may have to adjust the amounts of materials.*

---

1 Determine the location, height, size, and configuration of your deck.

Decide where you will attach the deck to your home. You will probably want to locate it beside a door, for easy access to the deck from the house. Don't hide or cover utility meters, buried utility lines, septic tanks, wells, spigots, drainage tiles — anything you may need to use, check, maintain, or repair from time to time.

Note: You can also build a freestanding deck, away from your home, by adding more support posts and railings. The same rule of thumb applies to its location: Don't build it over something you may need to access later.

Next, determine how far above the ground you need to build the deck. Normally, you'll want the deck surface to be ½″–1″ below the doorsill. (See Figure 1.) The bottom edge of the doorsill on the house shown is 29″ above ground level. The surface of the deck is ½″ below it.

Finally, decide the size of the deck. If you have a large yard, it can be almost as big as you care to make it. If you have a small yard, there will be constraints. Not only will the yard itself be a limiting factor, but local codes will probably require you to set the deck back at least 10 feet from the property line.

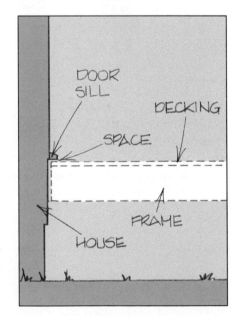

1/Plan to build your deck so the surface is just ½″–1″ below the bottom edge of the doorsill.

The location, height, and size of the deck determine its configuration. Sketch a rough plan on paper, or modify the plans shown. Carefully arrange the location of the supporting and nonsupporting posts, stairs, and railings.

Calculate the span of the beams and joists, then determine the sizes needed to build the deck, using the chart in "Selecting Materials," beginning on page 1. Revise the Materials List, changing the sizes and the amounts according to your plan. Then purchase the materials.

Note: As designed, this deck is supported by a pole foundation. However, decks are often built on piers and beams. If you substitute a different foundation, this will also alter the Materials List.

2 Lay out and set the posts.

Locate the supporting posts with stakes and string. Dig postholes, set the posts, and cut them off at the proper height. (Refer to "Step-by-Step: Setting Pole Foundations," page 45.)

3 Assemble the frame.

Cut the beams and the long 2 x 10 ledger. The beams attach to the ledger with butt joints — cut these ends square. Miter the other corners. Also cut ½"-thick wood spacers from scraps. These spacers will hold the ledger out slightly from the house wall, allowing air to circulate behind the ledger. The spacers prevent moisture from collecting here and damaging the house siding.

Locate the studs in the house wall against which you want to attach the deck. If the house is made of concrete blocks or masonry, sink expandable lead anchors in the wall. Attach the ledger to the wall with ⅜" x 6" lag screws. Drive the screws through the ledger and the spacers, and into the studs or lead anchors, as shown in the *Ledger-to-House Joinery Detail*.

Temporarily tack the beams in place on the posts with 16d common nails — don't drive the nails all the way. Make sure the top edge of each beam is level and the same distance from the tops of the supporting posts as the other beams. Using 16d spiral nails, join the mitered ends of the beams. Nail the square ends of the beams to the ends of the ledger.

Attach the beams to the supporting posts with ⅜" x 4" lag screws, driving the screws from the outside faces of the beams. Remove the 16d nails as you drive the screws. Cut the nonsupporting posts to length and screw them to the frame as well. Make sure that you attach all the posts to the *inside* surface of the frame, and that the posts are straight up and down.

4 Attach the short ledgers and joists.

Cut short ledgers to support the decking from 2 x 4 stock. Attach the short ledgers to the beams parallel to the joists *only* and fit them around the posts as shown in the *Ledger and Post Detail*. Carefully measure the decking to determine its thickness — this may be as little as 1" or as much as 1¼", depending on how the decking was milled. Position each ledger below the top edge of the beam, making sure that the distance is equal to the thickness of the decking. (See Figure 2.) Attach the ledgers to the frame with 10d spiral nails.

Measure and mark the location of the 2 x 8 joists — these should be spaced every 16", on center. At each mark, attach a joist hanger with roofing nails. Cut the joists to length, and place them in the hangers. The top of the joists should be even with the top of the short ledgers. Secure the joists in the hangers with roofing nails.

2/Attach the 2 x 4 ledgers to the beams, slightly below the top edges. Drive the nails at an angle so they don't come through the outside of the frame. You can also use lag screws to secure the ledgers, as shown.

5

Install the decking. Cut the decking to length. Where necessary, fit it around the posts, making the notches with a saber saw or hand saw. Attach the decking to the joists and the ledgers with 8d spiral nails, spacing the boards ¼"–½" apart, as shown in the *Top Elevation*. (See Figure 3.)

TRY THIS! Cut blocks of scrap wood, ¼"–½" thick, to help space the decking planks evenly. To position each successive plank, put several blocks against the edge of the plank you just installed. Slide another plank up against the blocks and nail it down. Remove the blocks and repeat.

3/When installed, the top surface of the deck must be flush with the top edge of the beam. *This hides the decking end grain, and gives the deck a more finished appearance.*

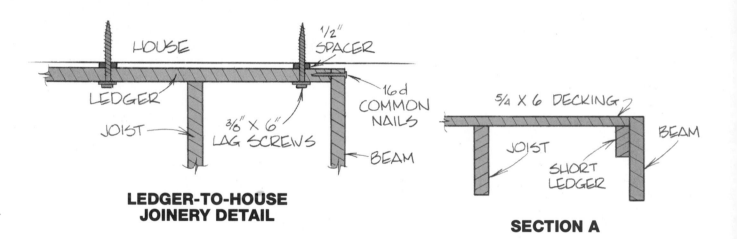

LEDGER-TO-HOUSE JOINERY DETAIL

HOUSE

½" SPACER

LEDGER

JOIST

⅜" X 6" LAG SCREWS

16d COMMON NAILS

BEAM

SECTION A

5/4 X 6 DECKING

JOIST

SHORT LEDGER

BEAM

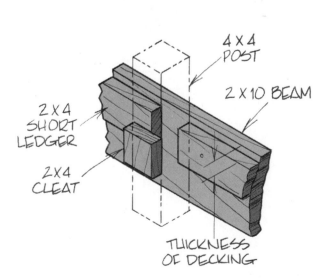

LEDGER/POST DETAIL

4 X 4 POST

2 X 10 BEAM

2 X 4 SHORT LEDGER

2 X 4 CLEAT

THICKNESS OF DECKING

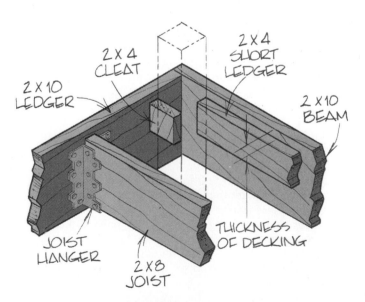

FRAMING DETAIL

2 X 4 CLEAT

2 X 10 LEDGER

2 X 4 SHORT LEDGER

2 X 10 BEAM

JOIST HANGER

2 X 8 JOIST

THICKNESS OF DECKING

6

Make and attach the stairs. As shown in the *Side Elevation*, the stair units for this deck each have a rise of 28½″ and a run of 31″. This may change depending on the landscape and how far above the ground you build the deck. If the ground is uneven, the rise and run may be different for each unit.

Measure the rise of each stair unit, then calculate the run and the number of treads. Cut the stair ledgers and attach them to the outside of the frame with 10d spiral nails. Then cut the risers and set the anchor blocks. Finally, attach the ledger, risers, and treads with 16d spiral nails. (Refer to "Step-by-Step: Building Stairs," page 48.)

7

Fit the rails to the posts. The rails are dadoed into the posts. To figure the length of the rails (including the tenons), carefully measure the distance between the posts and add 1½″ to each measurement. Cut both the top and bottom rails to size.

Cut a ¾″-wide, ⅜″-long tongue in both ends of each rail, using a table-mounted router or a dado cutter. Cut matching dadoes in the posts with a router. (See Figure 4.) Test fit the rails in place between the posts, but *don't* nail them.

4/Rout dadoes in the posts ¾″ wide, ¾″ deep, but only 2¾″ long. The outside end of each dado should be **blind,** as shown. Stop it ¾″ from the outside edge. Notch the outside edge of each tongue and round it to fit the blind dado.

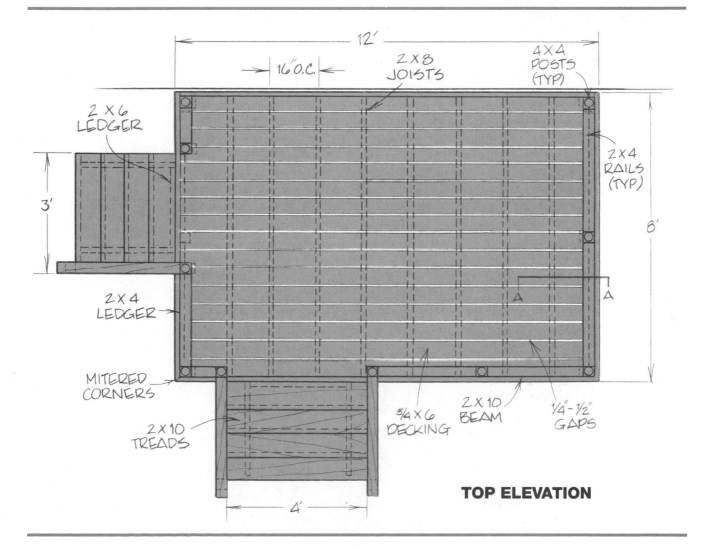

TOP ELEVATION

8

Cut and attach the spacers to the top rails. Measure and mark the positions of the balusters on the top rails *only*. Then slide the rails out from between the posts.

Cut baluster spacers from 1 x 6 stock. These spacers must fit in the 1½″-wide groove in the bottom face of each top rail, as shown in the *Top Rail/Baluster Detail*. (This groove is usually precut in handrail stock. If you can't get handrails with the grooves already cut, you'll have to cut your own.) Using 6d common nails, attach the spacers to the rails, leaving a 1½″ gap between each spacer.

9

Mortise the bottom rails. From scraps of hardboard and 2 x 4s, make a *Baluster Mortising Jig*, as shown. This jig is a template to help rout square mortises.

Lay each top rail over its matching bottom rail. The spacers on the bottom of the top rail should rest on the top face of the bottom rail. Wherever there's a gap between spacers, mark the bottom rail to indicate the position of the baluster mortises. (The spacers will protrude ⅜″–½″, so you can easily get a pencil between the two rails.)

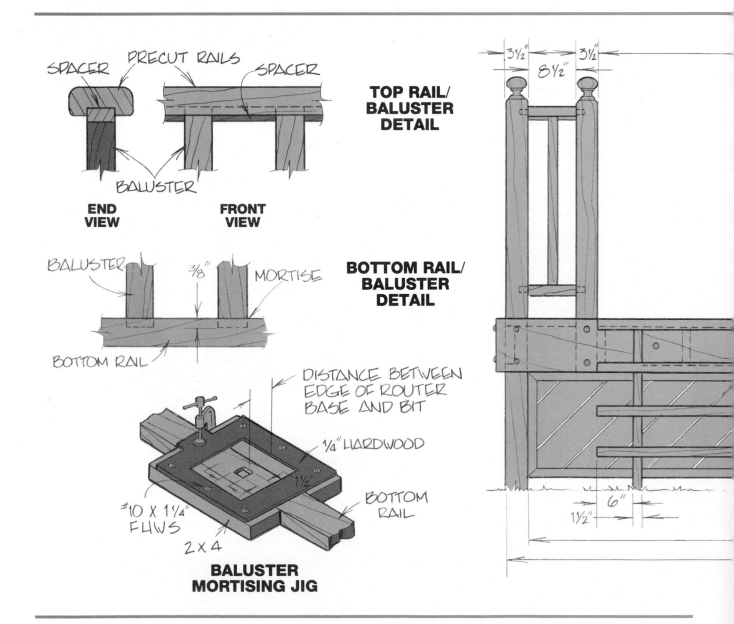

SPACER   PRECUT RAILS   SPACER

BALUSTER

END VIEW   **FRONT VIEW**

TOP RAIL/ BALUSTER DETAIL

BALUSTER   ⅜″   MORTISE

BOTTOM RAIL

BOTTOM RAIL/ BALUSTER DETAIL

DISTANCE BETWEEN EDGE OF ROUTER BASE AND BIT

¼″ HARDWOOD

1½″

BOTTOM RAIL

#10 X 1¼″ FHWS

2 X 4

BALUSTER MORTISING JIG

3½″   8½″   3½″

6″

1½″

Lay a bottom rail across two sawhorses with the marked side up, and place the jig over it. Align the jig with a baluster mark and clamp it to the rail. Rout the mortise with a straight bit, using the jig to guide the router. (See Figure 5.) The router will cut a 1½″-square mortise with rounded corners. Repeat until you have made all the mortises.

5/Cut the baluster mortises in the bottom rails with a router and a ¼″ straight bit. Guide the router with a jig, making each mortise precisely 1½″ wide, 1½″ long, and ⅜″ deep.

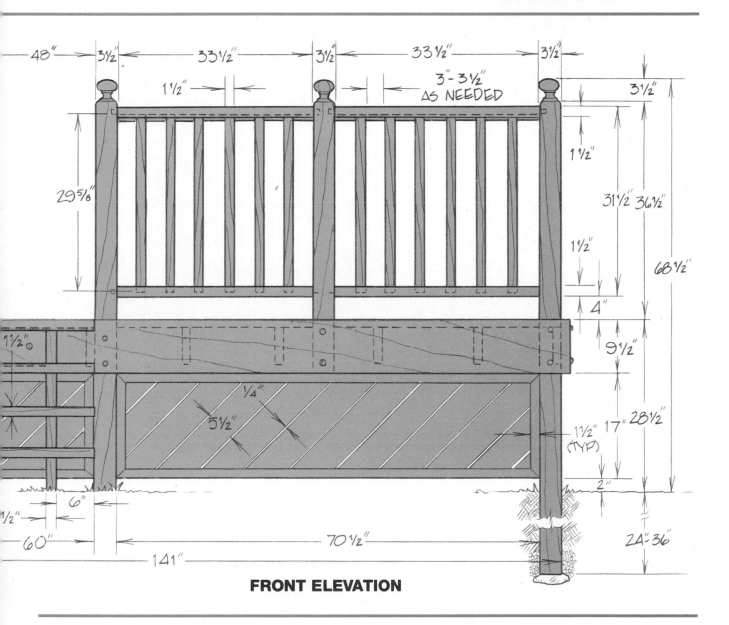

FRONT ELEVATION

10

Attach the rails and balusters to the posts. Cut the balusters to length from 2 x 2 stock, and fit them in the mortises. You shouldn't need to round the corners of the balusters to get them to fit — construction-grade 2 x 2 stock is already slightly rounded on the corners. However, you can round the corners a little more with a block plane if necessary.

Place each top rail over the baluster assemblies, fitting the top ends of the balusters between the spacers. Then slide the assembly between the proper posts, fitting the tongues into the dadoes. Secure the rails to the posts with 10d finishing nails. Drive the nails at an angle, down through the rails and into the posts. Set the heads of the nails below the surface of the wood.

11

Install the stair rails. Cut the 4 x 4 x 12' stock in two, making 6'-long posts. Set these posts in holes in the ground just beyond the stairs, on both ends of the treads. Line them up with the posts at the

top of the stairs, as shown in the *Side Elevation*. The posts at the top of the stairs are commonly referred to as *upper newel posts,* and those at the bottom are *lower newel posts.*

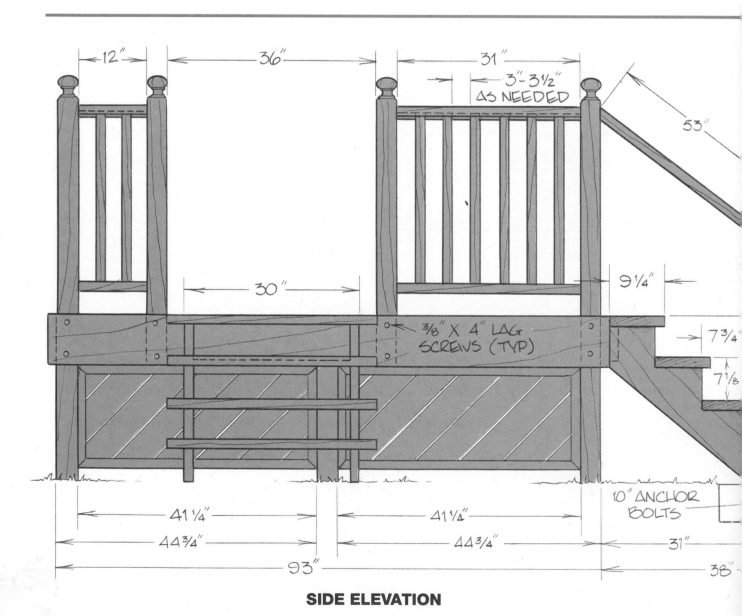

SIDE ELEVATION

Calculate the *angle of descent* for each stair unit. To do this, lay a 2 x 4 or a long, straight scrap along the front edges of the stair treads, as if it were a ramp going up the stairs. With a sliding T-bevel or a protractor, measure the angle between this board and the top surface of a tread. (See Figure 6.) In the example, the measured angle is 50°. Subtract this angle from 90° — 90° minus 50° is 40° — to calculate the angle of descent.

Use this angle to determine where to cut off the lower newel posts. To locate that angle on a lower newel post, drive a small nail into the upper post, even with the *bottom edge* of the top rail. Tie a string to the nail and stretch it taut between the posts, touching the inside faces. Have a helper move the free (lower) end of the string up and down while you measure the angle between the string and the upper post. When that

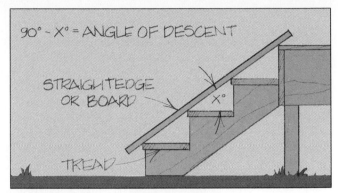

6/If you draw a diagonal between the rise and run of a stair, the slope of the diagonal is the angle of descent.

The easiest way to find this angle is to lean a board across the stair treads and measure the angle as shown.

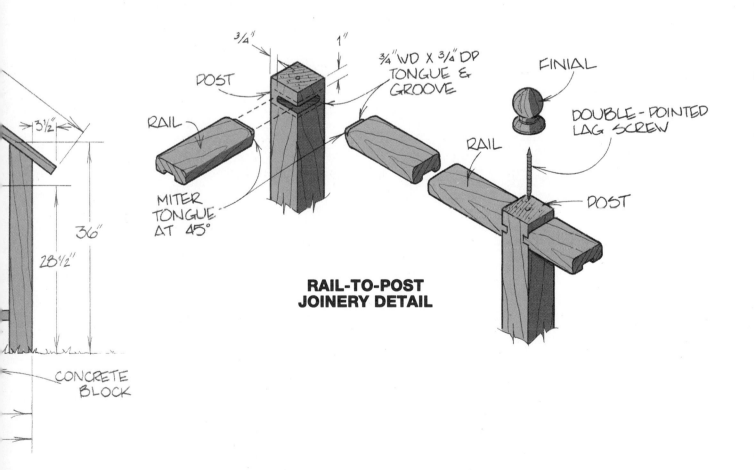

RAIL-TO-POST JOINERY DETAIL

angle is equal to the angle of descent, have the helper mark the position of the string on the lower post. (See Figure 7.)

Repeat this process for the other set of newel posts. With a string and a string level, check that your marks on both lower posts are even with each other. If not, repeat your measurements and move one or both marks so they are even. Cut off the posts at the marks, mitering the ends at the angle of descent.

Cut the handrails, mitering one end of each rail at the angle of descent. Put one rail in place with the mitered end against the upper newel post, even with the top rail. Mark the opposite (lower) end approximately 3½″ out from the lower post. Cut the lower end square. Use this rail to measure and mark the lower end of the other rail, and cut it square. Attach both rails to the posts with 16d spiral nails.

When all the rails are installed, attach finials to the tops of the posts with double-pointed lag screws.

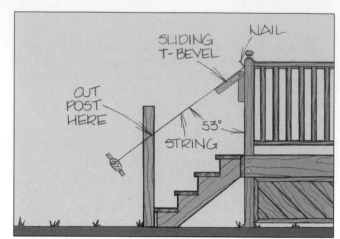

7/Use a string and a sliding T-bevel to determine where to cut off each lower newel post. Stretch the string between the upper and lower newel posts, as shown, and measure the angle between the string and the upper post with the bevel. Move the string up and down on the lower post until the string is at the angle of descent.

12 Assemble the skirts and attach them to the deck.

To determine the dimensions of the skirt panels, carefully measure the spaces between the support posts, and the distances from the ground to the lower edges of the deck frame between each set of posts. Each panel should be as wide as the distance between the posts, and 2″ shorter than the distance from the ground to the frame.

Build the skirt frames first. Rip 1½″-wide strips of decking and miter the corners, as shown in the *Skirt Panel Detail*. Assemble the frames with 10d finishing nails. Lay the frames in place between the posts to check their fit.

When you're satisfied that the frames fit, cover the back of each with decking. Secure the decking with 6d common nails. As shown in the drawings, this decking runs across the frames diagonally at 45°. This is not necessary; it simply creates a pleasing visual effect when the skirt panels are installed. You can also attach the decking vertically, horizontally, or at alternating angles — whatever your tastes dictate.

Position the skirt panels between the posts. The top edge of the panel frame must be flush with the bottom edge of the beam. Drive 16d common nails through the frame and into the posts. You may have to drill pilot holes for the nails to keep them from splitting the frame members.

At least one panel should be movable, in case you need to crawl under the deck or want to use the space. If you will only need to detach it *occasionally,* fasten it with #14 x 3½″ flathead screws instead of nails. If you will need to open it *often,* hinge the top edge to the beam and secure the bottom corners with sliding bolt catches.

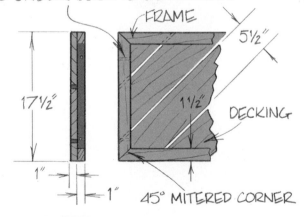

SECTION

SKIRT PANEL DETAIL

Step-by-Step: Setting Pole Foundations

Pole foundations are perhaps the most common, least expensive, and easiest way to marry an outdoor structure to the earth. A wide variety of projects, from birdhouses to fences to large utility buildings, can be supported by poles set in the ground.

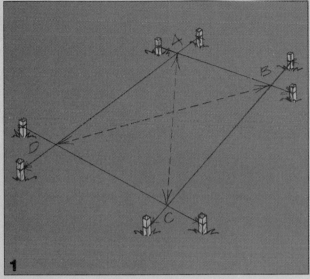

1

The first step in installing any foundation is to lay it out. To mark the location of the poles in the ground, drive several stakes as shown and stretch strings between them. You will set the posts where the strings cross. To be certain that the foundation is square, measure diagonally from corner to corner. The distance AC should equal BD. If it doesn't, move one or more of the stakes.

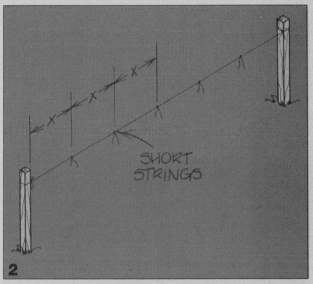

2

When laying out several poles in a straight line (for a fence or privacy screen, for example), simply stretch a string between two stakes. Measure along the string and mark the locations of the poles by tying on short lengths of string at these points.

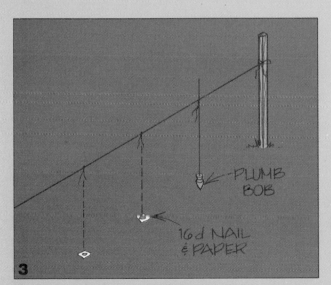

3

Use a plumb bob to transfer the location of each pole to the ground. Where the point of the bob touches the ground, press a 16d nail into the earth. To make the nail easier to see, drive it through a scrap of paper.

4

Remove the string. At each mark, dig a hole to bury the base of a pole. Each hole should be slightly larger in diameter than the pole (6"–8" for a 4 x 4 pole), and slightly deeper than the frost line (24"–36" in most areas). If you have a lot of holes to dig, you may want to rent a two-man power auger — this saves much time.

(Continued)

Step-by-Step: Setting Pole Foundations — Continued

5

Install the first poles at the four corners of the foundation (or the extreme ends of the fence row). Leave the posts long — you'll cut them off later. Place a large, flat rock or brick at the bottom of each hole to keep the pole from settling. Then put the pole in the hole, on top of the rock. Use a level to position the pole straight up and down, and hold it in place with stakes and braces.

6

If the area is particularly wet, you may want to coat the sides of the pole with roofing tar, 4"–6" above and below ground level. This will help prevent the pole from becoming water-logged where it meets the ground. **Don't coat the base of the pole!** This forms a "cup" of tar that prevents water from draining away from the pole — the pole becomes waterlogged and stays that way.

9

Adjust the row of posts so they just touch the string. Use the stakes and braces as levers to adjust the vertical position of the poles: Grasp a stake and move it back and forth until the pole is plumb; then tap the stake a little further into the ground so the pole stays put.

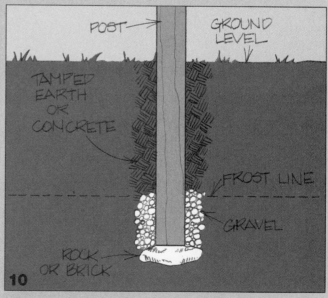

10

Fill in around the remaining poles with gravel and dirt. If necessary, use cement instead of dirt. When properly installed, the base of each pole should be below the frostline, resting on a rock, and surrounded by gravel. The pole should be held upright by tamped dirt or cement.

Throw a few shovelfuls of gravel in the hole around the base of the pole. This will help drain water away from the wood. Then fill the rest of the hole with earth. Tamp the earth down tightly, using a 2 x 4 or a heavy spud bar. If your structure is top heavy or the ground is very sandy and won't pack tightly, you may want to fill the hole with cement rather than tamped dirt. If you do, be careful not to set the base of the pole in cement. This creates a cement cup that collects water, and the pole rapidly becomes waterlogged.

Measure one of the corner (or end) poles and mark where you will cut it off. Put a small nail at the mark and tie a string to the nail. Stretch the string past the other posts, using a string level to make sure it is perfectly horizontal, and attach the other end to a nail in the far post.

Check that the string is still level. Then mark the tops of the posts where you will cut them off.

Remove the string and cut the tops of the posts level with each other, using a circular saw or hand saw. When you've finished, tamp the earth around the poles one more time and remove the braces.

Step-by-Step: Building Stairs

Most stairs must be custom-built to fit the structure and the landscape. These assemblies require a great deal of figuring, but they are not difficult to make if you plan them carefully.

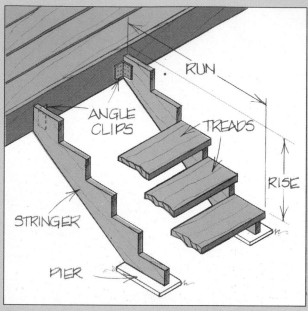

Stairs consist of one or more steps or **treads** laid across 2-by sawtooth-shaped boards called stringers. The upper ends of the **stringers** are attached to the deck or building with **angle clips,** while the lower ends rest on concrete **piers.** The height of the stairs, or the distance from the ground to the floor, is called the **rise,** and the depth, or distance from the structure to the outer edge of the first tread, is the **run.** Traditionally, most steps are 7″–9″ high and 11″–12″ deep. This makes a safe, comfortable rise and run.

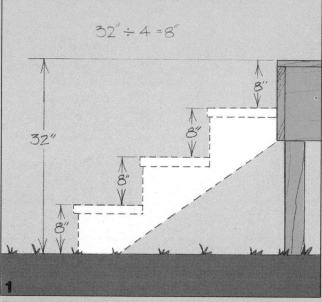

To design your steps, first figure out how many treads you need. Measure the rise and divide by **whole numbers** until you get an answer between 7 and 9. For example, if your rise is 32″, you might first divide it by 5. The answer 6⅖″, slightly less than needed for a comfortable stair height. Go down one whole number and try 4. A rise of 32″ divided by 4 is 8″ — just right. You need four stair treads, each one rising 8″ above the next. If the surface of the floor or deck will serve as the top tread (as it does in the following photos), you can subtract one tread and build a stair unit with just three steps.

Next, calculate the run. Multiply the number of treads times the ideal width of each tread — 11″–12″. On the example shown, the width is 11″. Three treads times 11″ makes a run of 33″ — the stairs will begin 33″ out from the structure. Lay out the stringers on 2 x 12 stock with a framing square: Lay the square on the board so the 8″ mark on the small (rise) arm and the 11″ mark on the large (run) arm are even with the top edge of the board. Trace the arms of the square, then move it down the board. Line up the small arm with the front end of the run line that you just drew, and repeat. Continue until you have marked as many steps as needed. To lay out the top end of the stringer, extend the **top riser** line **down** until it intersects the back edge of the stringer. This is the top cutoff line. Also extend the **bottom run** line **back** to the same edge. To lay out the bottom end of the stringer, measure **up** from the bottom run line the thickness of the tread stock. (If you're using 2-by stock, measure up 1½″.) At this point, draw a line parallel to the bottom run line — this is the bottom cutoff line.

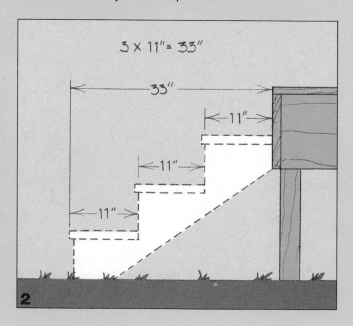

3 **Cut the first stringer** with a circular saw, saber saw, or hand saw. Use this as a template to mark other stringers, and cut as many as you need. Space the stringers the same as floor joists — every 16" or less, on center.

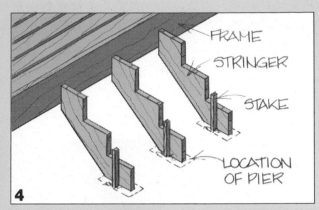

FRAME
STRINGER
STAKE
LOCATION OF PIER

4 **Set the stringers in place** temporarily, and put stakes in the ground where you will put the piers. Each stringer must rest on a pier.

5 **Set the piers in the ground,** making sure they're level with each other. The piers shown are ordinary concrete blocks. Bury each block in the ground so about 1" protrudes above the surface, and fill the cavities with cement. Wait until the cement begins to harden, then set a ½"-diameter anchor bolt in the front cavity. The anchor bolt should stick up 3"–4".

6 **After the cement hardens** completely, put the stringers back in place. Trace the position of each anchor bolt on its respective stringer. Drill a ¾"-diameter, 4"-deep hole in the bottom of each stringer, so the stringer will fit over its anchor bolt.

7 **Place several ½"-diameter flat washers** over each anchor bolt to keep the stringers from direct contact with the piers. Place the bottom end of each stringer over its anchor bolt, and attach the top end to the structure with angle clips. The top ends and the front edges of the stringers must all be even with each other.

8 **Attach the treads to the stringers** with 16d spiral nails. On the stair unit shown, the builder made the treads from 2 x 6 stock. Since each 2 x 6 is 5½" wide, he used two boards per tread — there was no need to rip special stock. The boards are spaced ½" apart, and the outer board overhangs each step by ½".

Custom-Built Birdhouses

Birds make any backyard a more pleasant place. Their songs are restful and soothing; watching them can be entertaining and educational. They're especially interesting if you can study them while nesting — tending the eggs, feeding the young, teaching nestlings to fly. And birds help control pests. Some species eat up to two thousand insects per bird per day!

One of the best ways to attract birds to your backyard is to build birdhouses. This is not the simple, grade-school project you may think it to be — birdhouse building is an exact science. Every species has its own nesting habits and prefers one of three basic types of home — a nesting shelf, a mounted birdhouse, or a hanging birdhouse. You can attract particular species by changing the opening size, floor size, ceiling height, distance above the ground, and location of the house.

The birdhouses shown are designed so you can easily adapt them to whatever species you wish. They all provide good drainage, ventilation, and easy access. The closed structures — the mounted and hanging birdhouses — can be easily opened for cleaning at the end of each nesting season.

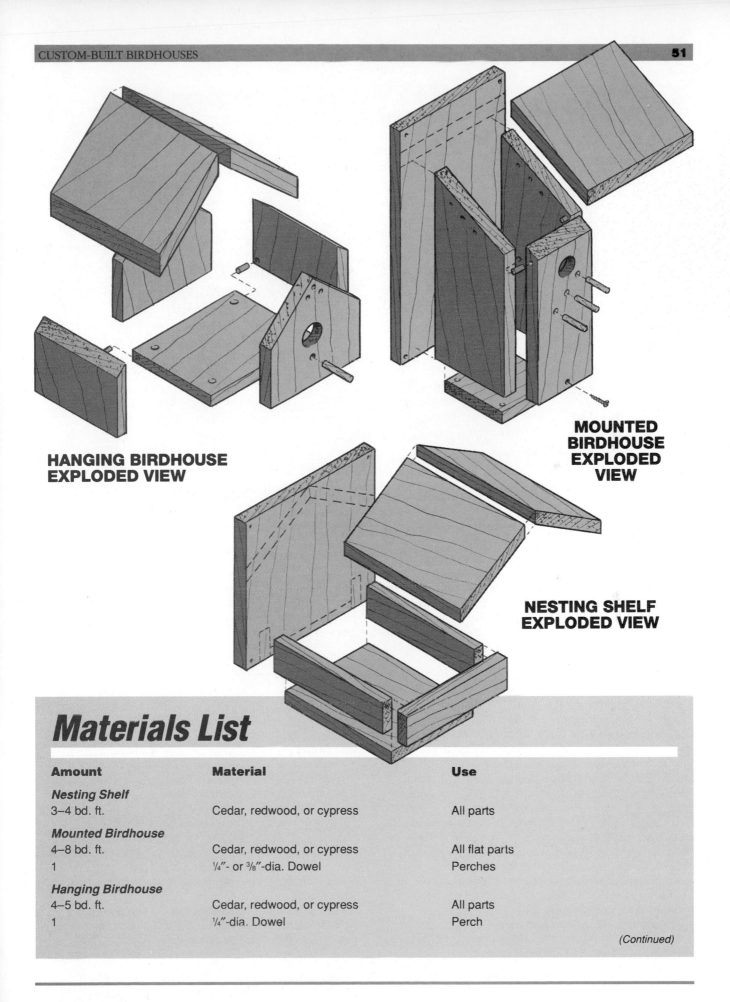

HANGING BIRDHOUSE EXPLODED VIEW

MOUNTED BIRDHOUSE EXPLODED VIEW

NESTING SHELF EXPLODED VIEW

Materials List

Amount	Material	Use
Nesting Shelf		
3–4 bd. ft.	Cedar, redwood, or cypress	All parts
Mounted Birdhouse		
4–8 bd. ft.	Cedar, redwood, or cypress	All flat parts
1	¼″- or ⅜″-dia. Dowel	Perches
Hanging Birdhouse		
4–5 bd. ft.	Cedar, redwood, or cypress	All parts
1	¼″-dia. Dowel	Perch

(Continued)

Materials List — Continued

Amount	Material	Use
HARDWARE		
Nesting Shelf		
1 pt.	Resorcinol or epoxy glue	Assemble roof halves and shelf
4–6	6d Finishing nails (galvanized)	Reinforce glue joints
8	#10 x 2″ Flathead wood screws	Attach roof and shelf to back
Mounted Birdhouse		
1 pt.	Resorcinol or epoxy glue	Assemble back, sides, roof
24–30	6d Finishing nails (galvanized)	Reinforce glue joints
2	¼″ dia. x 1″ Brass pivots	Mount front to sides
1	#10 x 1½″ Flathead wood screw (brass)	Secure front to bottom
Hanging Birdhouse		
1 pt.	Resorcinol or epoxy glue	Assemble roof halves, front, back, sides
24–30	6d Finishing nails (galvanized)	Reinforce glue joints
2	¼″ dia. x 1″ Brass pivots	Mount bottom to sides
1	#10 x 1½″ Flathead wood screw (brass)	Secure bottom to front
2	½″ Eye screws	Hang birdhouse

Note: *The amount of lumber required for a birdhouse will change according to the species you want to attract.*

1 Choose the species you want to attract.

Decide what species of birds you want to attract to your yard. If you're unsure, visit the library and consult bird-watching field guides to see what species are common to your area. You can also call your local chapter of the Audubon Society for information. Investigate a bird's preferred habitats, food sources, and nesting habits to see if your neighborhood offers the proper environment.

2 Choose the type of birdhouse you want to build.

Having decided the species and knowing its behavior, pick the *type* of birdhouse that will attract that bird — nesting shelf, mounted birdhouse, or hanging birdhouse. The vast majority of birds prefer closed-in, protected places to raise their young, and most of these prefer that their nests be securely *mounted* to a tree, pole, or building. Only the wren family likes to nest in a house that hangs by a rope or chain, swaying in the breeze. There are also a few species that won't nest in a traditional, closed-in birdhouse. Robins, swallows, and phoebes usually build their nests on open ledges or shelves.

There are several other factors you should consider:

Ventilation — Birdhouses must be well ventilated so they don't collect hot air. On a hot day, without enough airholes, newly-hatched young can die of dehydration and overexposure.

Drainage — Birdhouses must quickly drain away any rainwater that blows in. Young birds who can't yet leave their nest may drown if water collects in the bottom of the house.

Cleaning — Birds don't often use the same nest twice. After one brood has left, you can attract other birds to your birdhouse by cleaning it out and removing the old nesting materials. Build the house so you can open it easily, and try to mount it in a location that's not too difficult to reach.

All the designs shown offer good ventilation and drainage, and can be easily cleaned. If you modify the designs, make sure you retain these features.

3 *Design the birdhouse for the bird.* Not only do different species prefer different types of birdhouses, they also prefer a particular floor size. The ceiling should be a certain height, the entrance the proper diameter and height above the floor. Finally, the birdhouse should be mounted at a particular height above the ground.

Sketch the type of birdhouse you want to build, adapting the design according to the following chart:

Dimensions and Locations of Birdhouses

Species	Floor Size	Ceiling Height	Diameter of Entrance	Height of Entrance*	Height Above Ground
Nesting Shelves					
American Robin	6″ x 8″	8″	N/A	N/A	6′–15′
Barn Swallow	6″ x 6″	6″	N/A	N/A	8′–12′
Eastern Phoebe	6″ x 6″	6″	N/A	N/A	8′–12′
Song Sparrow	6″ x 6″	6″	N/A	N/A	1′–3′
Mounted Birdhouses					
American Kestrel	8″ x 8″	12″–15″	3″	9″–12″	10′–30′
Eastern Bluebird	5″ x 5″	8″	1½″	6″	5′–10′
Chickadee	4″ x 4″	8″–10″	1⅛″	6″–8″	6′–15′
Downy Woodpecker	4″ x 4″	8″–10″	1¼″	6″–8″	6′–20′
House Finch	6″ x 6″	6″	2″	4″	8′–12′
Northern Flicker	7″ x 7″	16″–18″	2½″	14″–16″	6′–20′
Purple Martin**	6″ x 6″	6″	2½″	1″	12′–20′
Nuthatch	4″ x 4″	8″–10″	1¼″	6″–8″	12′–20′
Red-bellied Woodpecker	6″ x 6″	12″–15″	2½″	9″–12″	12′–20′
Red-headed Woodpecker	6″ x 6″	12″–15″	2″	9″ 12″	12′–20′
Screech Owl	8″ x 8″	12″–15″	3″	9″–12″	10′–30′
Starling	6″ x 6″	16″–18″	2″	14″–16″	10′–25′
Titmouse	4″ x 4″	8″–10″	1¼″	6″–8″	6′–15′
Tree Swallow	5″ x 5″	6″	1½″	1″–5″	10′–15′
Hanging Birdhouses					
Carolina Wren	4 ″x 4″	6″–8″	1½″	4″–6″	6′–10′
House Wren	4″ x 4″	6″–8″	1″–1¼″	4″–6″	6′–10′
Winter Wren	4″ x 4″	6″–8″	1″–1¼″	4″–6″	6′–10′

*Distance from floor to entrance hole
**Purple martins prefer to live in colonies of 6 or more nesting pairs. Build a large mounted birdhouse with several 6″ x 6″ x 6″ compartments.

This chart was condensed from *Home for Birds,* published by the U.S Department of the Interior, Fish and Wildlife Service. You can obtain a more complete list from your local library or by writing:

Government Printing Office
North Capitol & H Streets
Washington, DC 20401

You can also call the Publications Department of the Fish and Wildlife Service at (703) 358-1711.

4 **Choose the materials.** Estimate the amount of materials needed to build your birdhouse, and purchase the lumber and hardware. When selecting lumber, choose a weather-resistant, *untreated* wood such as cedar, redwood, or cypress. Don't use pressure-treated lumber. Although this lumber is safe for most outdoor building projects, you should avoid it when the people (or animals) using the project will be in constant contact with the wood. The

chemicals used in pressure-treating are toxic, and birds are very susceptible to toxins.

Also give some thought to the hardware: Most birds shy away from bright, reflective materials. This won't be a problem if you purchase brass parts, because they quickly tarnish in the weather. However, if you use galvanized metal or stainless steel, you should paint it a flat black, brown, or green.

Building a Nesting Shelf

1 **Cut the parts to size.** Plane the wood ¾" thick. (If you want your birdhouse to have a rustic look, leave one side rough and use this on the *outside*.) Cut the pieces to the sizes needed. Bevel the edges of

the roof halves at 30°, and chamfer the corners of the bottom at 45°, as shown in the *Nesting Shelf/Front View* and *Nesting Shelf/Bottom Layout*.

2 **Drill mounting holes in the back.** Drill ¼"-diameter holes through the back, where shown in the *Nesting Shelf/Front View*. After you com-

plete the project, you can use these holes to mount the birdhouse to a building or tree.

3 **Assemble the nesting shelf.** Assemble the front, sides, and bottom with waterproof resorcinol or epoxy glue. Reinforce the glue joints with 6d finishing nails. Assemble the halves of the roof in the same manner, but don't try to hammer the nails. Because the roof halves meet at an awkward angle, you'll have to *spin* the nails into the wood. (See Figure 1.) Drive the nails from both right and left, to strengthen the roof as much as possible.

When the glue dries, attach the roof and shelf assemblies to the back with #10 x 2" flathead wood screws. Drive the screws through the back and into the roof, sides, and bottom.

1/To "spin" a finishing nail, mount it in a drill chuck. Turn on the drill and drive the nail into the wood, as if you were drilling with a small bit. When you have driven the nail as far as possible with the drill, loosen the chuck. Drive the nail home with a hammer and a nail set.

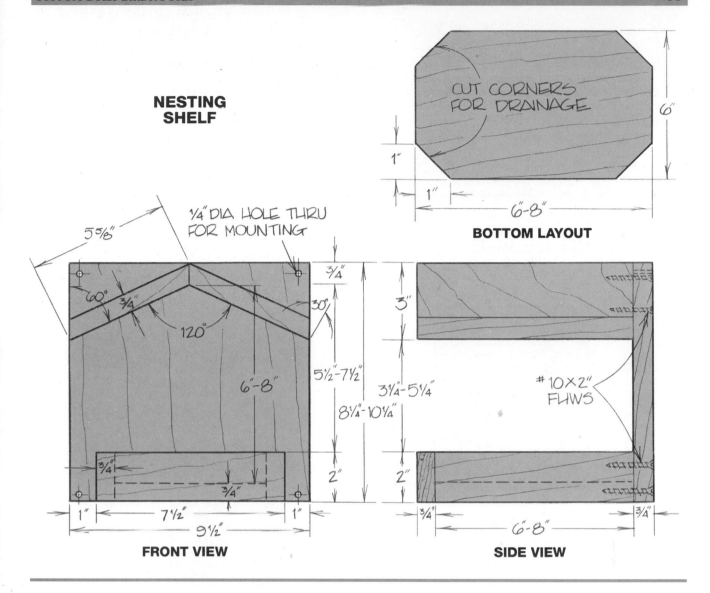

NESTING SHELF

5⅝"

¼" DIA HOLE THRU FOR MOUNTING

60°

¾"

120°

30°

¾"

6–8"

¾"

¾"

1"

7½"

1"

9½"

FRONT VIEW

¾"

5½–7½"

8¼–10¼"

2"

CUT CORNERS FOR DRAINAGE

1"

6"

1"

6–8"

BOTTOM LAYOUT

3"

3¼–5¼"

2"

#10×2" FHWS

¾"

6–8"

¾"

SIDE VIEW

Building a Mounted Birdhouse

1 **Cut the parts to size.** Plane the wood ¾" thick, and cut the pieces to the sizes needed. Bevel *both* edges of the roof and miter the *top* edge of the front at 60°, as shown in the *Mounted Birdhouse/Side View*. Cut the corners of the bottom at 45°, as shown in the *Mounted Birdhouse/Bottom Layout*.

2 **Drill the necessary holes.** With a hole-saw or a large multispur bit, cut the entrance hole in the front. Then drill the other holes needed for mounting, ventilation, perches, etc.:

■ ½"-diameter holes through the sides, near the upper edge, for ventilation

■ ⅜"- or ¼"-diameter holes through the front, just below the entrance hole, to mount the perches

■ ¼"-diameter holes through the back, near the corners, to mount the birdhouse

■ ¼"-diameter, ½"-deep holes in the edges of top and the inside faces of the sides, to hold the pivots

3

Assemble the birdhouse. From a ¼″-diameter brass rod, cut two 1″-long pivot pins. (You can purchase brass rod from a welding supplier.) Place a pin in each front pivot hole. Put the sides in place, fitting the side pivot holes over the brass pins. Assemble the sides, bottom, back, and roof with waterproof resorcinol or epoxy glue, and reinforce the glue joints with 6d finishing nails. Be careful not to get

any glue on the front — the front should swing freely on the pivot pins.

Glue the perches in the front. To keep the front from swinging, drive a #10 x 1½″ flathead wood screw through it and into the bottom. When you need to clean the birdhouse, loosen the screw and swing the front forward.

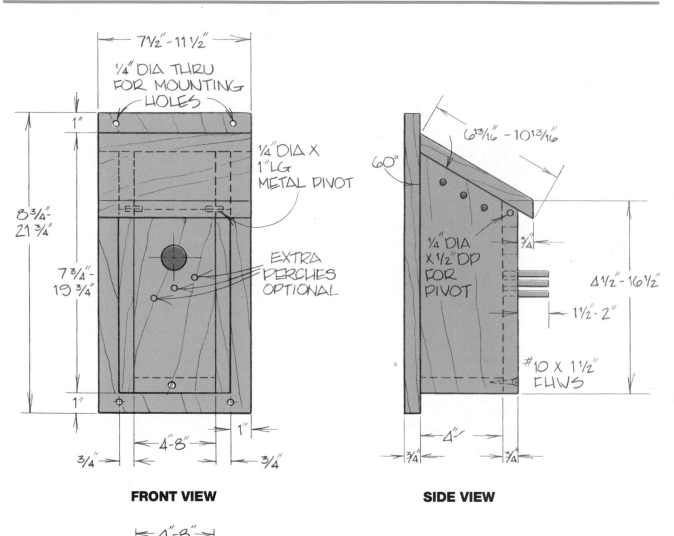

FRONT VIEW

SIDE VIEW

BOTTOM LAYOUT

MOUNTED BIRDHOUSE

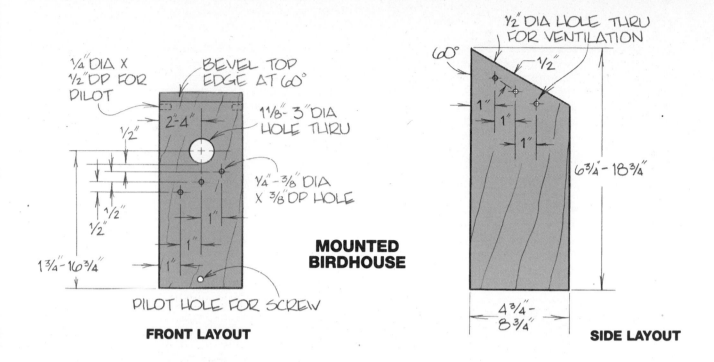

MOUNTED BIRDHOUSE

¼" DIA X ½" DP FOR PILOT

BEVEL TOP EDGE AT 60°

2"-4"

½"

1⅛"- 3" DIA HOLE THRU

¼"- ⅜" DIA X ⅜" DP HOLE

½"

½"

1"

1"

1"

1¾"- 16¾"

PILOT HOLE FOR SCREW

FRONT LAYOUT

½" DIA HOLE THRU FOR VENTILATION

60°

½"

1"

1"

1"

6¾" - 18¾"

4¾" - 8¾"

SIDE LAYOUT

Building a Hanging Birdhouse

1 **Cut the parts to size.** Plane the wood ¾" thick, and cut the pieces to the sizes needed. Bevel *both* edges of the roof halves and the *top* edges of the sides at 45°, as shown in the *Hanging Birdhouse/ Front View.* Cut a gable shape in the front and back, mitering the top end of each board twice at 45°, to form a point as shown in the *Hanging Birdhouse/Front Layout.*

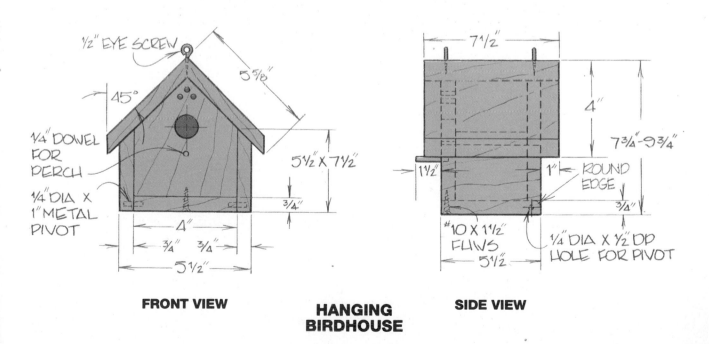

½" EYE SCREW

45°

5⅝"

¼" DOWEL FOR PERCH

¼" DIA X 1" METAL PIVOT

4"

¾"   ¾"

5½"

5½" X 7½"

¾"

FRONT VIEW

7½"

4"

7¾"-9¾"

1½"

1" ROUND EDGE

¾"

#10 X 1½" FHWS

5½"

¼" DIA X ½" DP HOLE FOR PIVOT

SIDE VIEW

HANGING BIRDHOUSE

2 Drill the necessary holes.

Drill the necessary holes. With a hole-saw or a large multispur bit, cut the entrance hole in the front. Then drill the other holes needed for ventilation, pivots, etc.:

■ ½"-diameter holes through the bottom, as shown in the *Hanging Birdhouse/Bottom Layout,* for drainage

■ ⅜"-diameter holes through the front and back, near the gable points, for ventilation

■ ¼"-diameter hole through the front, just below the entrance hole, to mount the perch

■ ¼"-diameter, ½"-deep holes in the edges of bottom and the inside faces of the sides, to hold the pivots

3 Assemble the birdhouse.

Assemble the birdhouse. Using a hand plane or a rasp, round the back edge of the bottom, as shown in the *Hanging Birdhouse/Side View.* From a ¼"-diameter brass rod, cut two 1"-long pivot pins, and place a pin in each bottom pivot hole. Put the sides in place, fitting the side pivot holes over the pivots. Assemble the sides, front, back, and roof halves with waterproof resorcinol or epoxy glue, and reinforce the glue joints with 6d finishing nails. Be careful not to get any glue on the bottom — it should swing freely on the pivots.

Glue the perch in the front of the birdhouse, and install eye screws at the roof peak. To keep the bottom from swinging, drive a #10 x 1½" flathead wood screw up through it and into the front. When you need to clean the birdhouse, loosen the screw and let the bottom swing down.

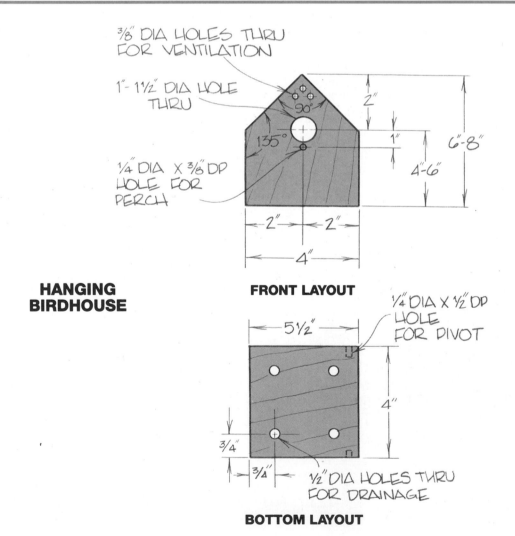

HANGING BIRDHOUSE

FRONT LAYOUT

BOTTOM LAYOUT

Mounting and Hanging the Birdhouse

To attach a birdhouse or nesting shelf to a vertical surface such as a tree, building, or fence, drive roundhead screws through the back and into the support. If possible, try to place the house where you can reach it without too much trouble when it's time to clean it. Also, try to position the house so it tilts forward *slightly,* at 5°–10°. Most birds swoop up to a perch or entrance, and this angle helps facilitate an easy, comfortable landing.

To mount a birdhouse on a pole, first attach a pipe flange to the top of the pole. Drive screws through the flange, into the bottom of the house. (See Figure 2.) Set the bottom of the pole in a concrete base, cast into the ground. (See Figure 3.) When you need to clean the house, remove the pole from the base and lower the house.

TRY THIS! Unless the birdhouse is very heavy or needs to be mounted high above the ground, you can use a rigid PVC pipe and a plastic flange to support it. This saves a lot of weight, making it easier to raise and lower the house.

To hang a birdhouse, attach light chains to a tree limb, under the eaves of a roof, or to some other suitable location. Try to find a place that's well protected from the wind — the birdhouse shouldn't sway constantly. Fasten the chains to the birdhouse with S-hooks. (See Figure 4.)

One last thought: Although some people take their birdhouses inside for winter storage, consider leaving yours out — particularly if you live in a cold climate. You may attract wintertime birds. If you've built a nesting shelf, use it as a wintertime tray feeder. If the house is enclosed, leave it out as a shelter. Birds that ordinarily wouldn't nest in the birdhouse during the summer will huddle together inside for protection against the winter cold.

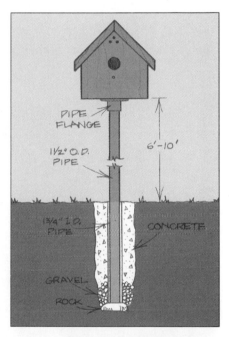

3/To mount the pole in the ground, dig a hole 24"–36" deep. Cut a length of iron pipe with an **inside** diameter slightly larger than the **outside** diameter of the pole. Place the pipe in the hole and carefully fill the hole with concrete **but not the pipe.** When the concrete sets, insert the bottom end of the pole into the pipe.

2/To mount a birdhouse on a pole, first mount a pipe flange on the upper end of the pole. Screw the flange to the bottom of the birdhouse.

4/To hang the birdhouse, drive two screw hooks into the underside of a tree limb or roof eaves, 6"–8" apart. Place the ends of two light chains over the screw hooks — one on each hook — and then attach the other ends to the birdhouse with S-hooks.

Sunshade

While everyone loves a warm, sunny day, too much heat and too much sun can make a deck or backyard uncomfortable — or even painful or dangerous. Overexposure to sunlight can cause heat stroke, sun poisoning, and skin cancer.

If you like to spend long periods outdoors on sunny days, it's wise to build a sunshade. This simple structure provides protection from direct sunlight. The closely spaced slats on this rooflike frame diffuse the sunlight, but don't block it completely. Sunshine filtered through a sunshade is not as intense, but it's just as enjoyable.

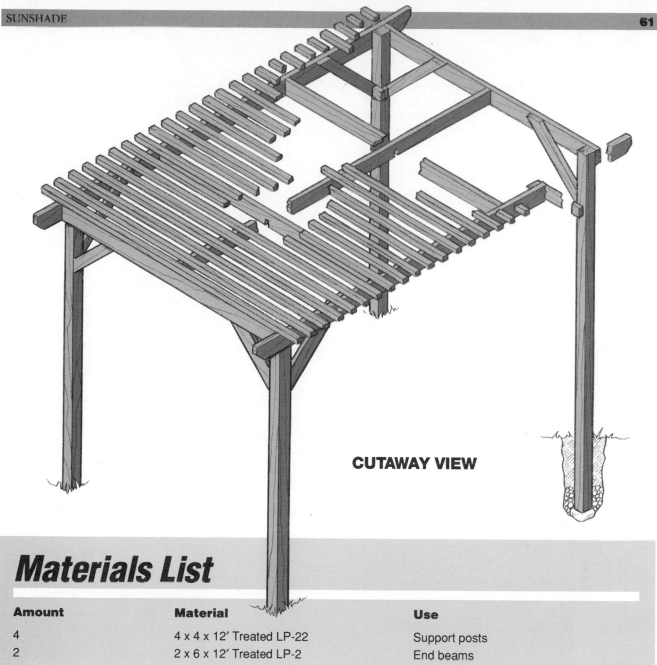

CUTAWAY VIEW

Materials List

Amount	Material	Use
4	4 x 4 x 12′ Treated LP-22	Support posts
2	2 x 6 x 12′ Treated LP-2	End beams
1	2 x 6 x 10′ Treated LP-2	Middle beam
4	2 x 6 x 8′ Treated LP-2	Rafters
5	2 x 4 x 8′ Treated LP-2	Braces, cleats
36–40	2 x 2 x 10′ Treated LP-2	Sunshade slats

HARDWARE

12	³⁄₈″ x 5½″ Carriage bolts, washers, and nuts	Attach sunshade frame to posts
16	³⁄₈″ x 3½″ Carriage bolts, washers, and nuts	Attach braces to frame
2 lbs.	16d Common nails (galvanized)	Assemble frame, attach braces to posts
5 lbs.	10d Spiral nails (galvanized)	Attach slats to frame

Note: *This list was figured for the project **as shown in the drawings**. If you change the size or configuration of the sunshade, you may have to adjust the amounts of materials.*

1 Calculate the spacing between the slats. The closer you live to the equator, the more intense the sun will be. And the more intense the sun, the more protection you need from it. To provide more protection, space the slats on your sunshade closer together.

Calculate the distance between the slats using your latitude. Look at any map or globe and find the latitude line that runs east-to-west through your area. If you live in the contiguous United States, your latitude will be somewhere between 25° and 49°.

To calculate the spacing of the slats, create a small side view of just one slat on your sunshade. To do this, draw a straight, horizontal line on a piece of paper to represent the top edge of the sunshade frame. Then draw a square, 1½″ to a side, on top of this frame to represent the full-size end view of a slat.

On a year-round average, sunlight will strike the slats on your sunshade at an angle equal to your latitude. Draw a diagonal line at that angle (your latitude) from an upper corner of the slat down to the top of the frame. On a typical sunny day, everything inside that angle will be in the shade; everything outside will be in the sun. For the sunshade to cast complete shade — and diffuse the sunlight completely — you should put another slat where the diagonal line meets the frame.

The proper spacing between the slats, then, is equal to the distance from the bottom corner of the first slat to the point where the diagonal line meets the frame. (See Figure 1.)

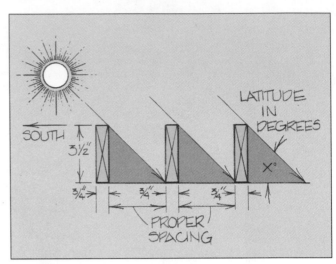

1/Use the **degree of latitude** for your locale to figure the spacing between slats. The smaller the latitude, the closer you should space the slats.

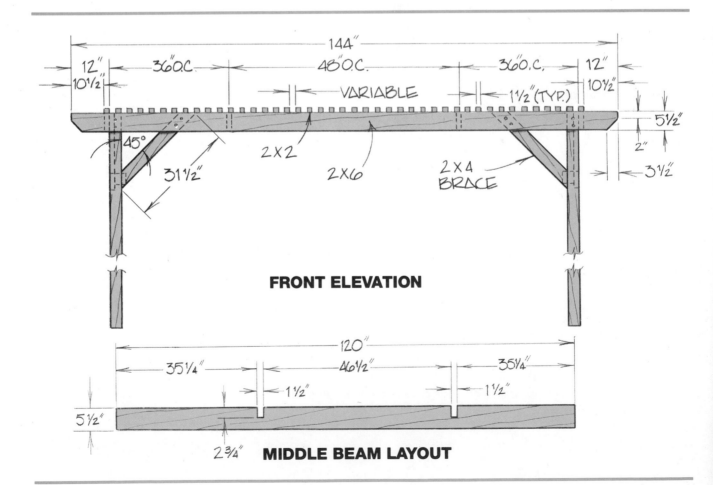

FRONT ELEVATION

MIDDLE BEAM LAYOUT

2

Determine the size and configuration of the sunshade. Measure the size of the area you want to cover to determine the size of the sunshade. Sketch a simple plan of the project or change the dimensions on the drawings shown. Consider the span of the beams and joists — if the span is large enough, you may have to add several support posts to the structure, or increase the size of the beams and rafters.

Note: On this particular project, the roofing materials — the slats — are much lighter than normal roofing, and they will collect little rainwater or snow. Because of this, you can span slightly longer distances with a rafter or beam than you would ordinarily. Unless there will be some other strain on the sunshade, you can safely increase a span by 25 percent over what is shown in the "Joist Size" and "Rafter Size" charts on page 3.

Once you know the space between the slats and the lumber sizes, calculate the amounts you'll need. Purchase the materials.

3

Set the posts. Lay out the posts and mark their locations with stakes. Dig the postholes, and set the posts in the holes. Fill in the holes and cut the poles to the proper height. (Refer to "Step-by-Step: Setting Pole Foundations," page 45.)

TRY THIS!
You can also attach the post directly to the deck or patio using post anchors.

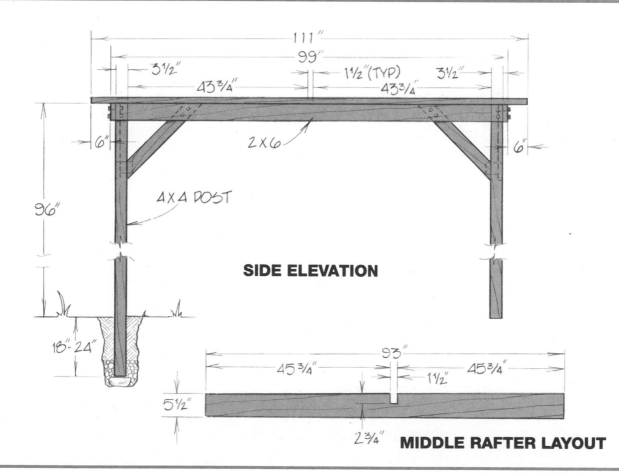

SIDE ELEVATION

MIDDLE RAFTER LAYOUT

4 **Assemble the roof frame.** Cut the beams and rafters to the lengths shown in the *Roof Frame Layout*. Also cut the braces, mitering the ends at 45°. Chamfer the corners of the two outside beams, as shown in the *Front Elevation*. Notch the middle beam and the middle rafters as shown in the *Middle Beam Layout* and *Middle Rafter Layout*. (See Figure 2.)

With 16d nails, temporarily attach the outside beams and rafters to the posts. Don't drive the nails home — you'll need to pull them out later. Drill ⅜"-diameter holes through the beams, rafters, and posts, as shown in the *Frame-to-Post Joinery Detail*. Secure the parts with ⅜" x 5½" carriage bolts, and remove the nails.

Attach the middle beam between the outside rafters with 16d nails. Then lay the middle rafters in place, fitting the notches in the rafters over those in the beam, as shown in the *Frame Joinery Detail*. Nail the middle rafters to the outside beams.

2/To make each notch, first cut the sides with a saber saw or a hand saw. Then separate the waste with a chisel.

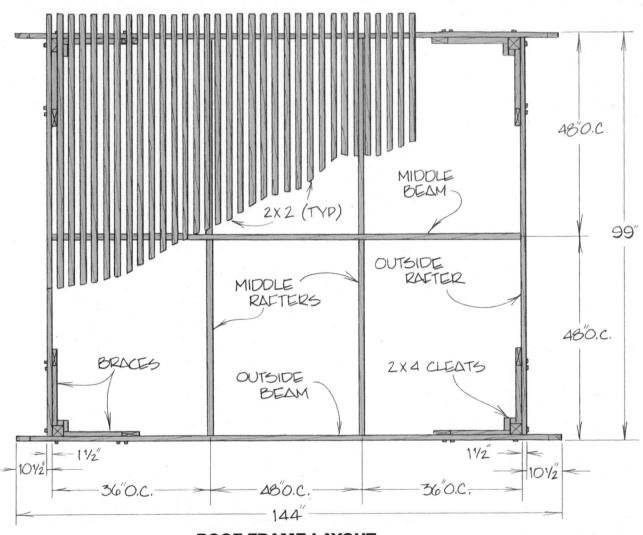

ROOF FRAME LAYOUT

5

Attach the braces. Clamp each brace in place on an outside beam or rafter. Make sure the lower end is flush against the post, then drill ⅜″-diameter holes through the sunshade frame and the upper end. Secure the upper end of the brace with carriage bolts.

With 16d nails, attach 2 x 4 cleats to a post, beside the lower end of the braces. Nail the braces into the cleats, as shown in the *Frame-to-Post Joinery Detail*. (See Figure 3.) Repeat for the remaining posts and braces.

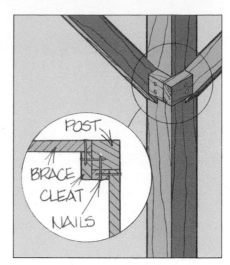

3/To attach the lower end of the braces to the posts, first nail cleats to the posts. Then nail the braces to the cleats.

6

Attach the slats to the sunshade. Cut the slats so they overhang the frame by 6″ on either end. Place each slat on the frame, spacing it the proper distance from the previous slat. Attach the slat with 10d spiral nails.

TRY THIS! To space all the slats evenly, cut several blocks of wood to the same thickness as the spacing you want to maintain. After nailing a slat to the frame, place the blocks on the beams, against the slat. Put another slat on the frame, slide it against the blocks, and nail it in place. Remove the blocks from between the slats and repeat.

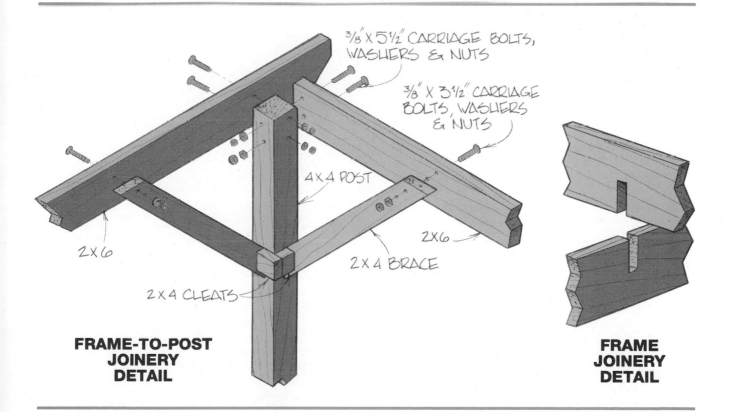

FRAME-TO-POST JOINERY DETAIL

FRAME JOINERY DETAIL

Step-by-Step: Pouring a Concrete Pad

Concrete pads are the most time-consuming foundations to install, but they have many advantages. They can be installed almost level with the ground, so you don't have to make stairs or a ramp to enter and exit the building. They are more permanent than most other types of foundations, and well suited to all kinds of soils and climates. In some applications, they may be less expensive than pole or pier-and-beam foundations, since they don't require any wood. Finally, the concrete surface is harder, stronger, and more durable than wooden flooring.

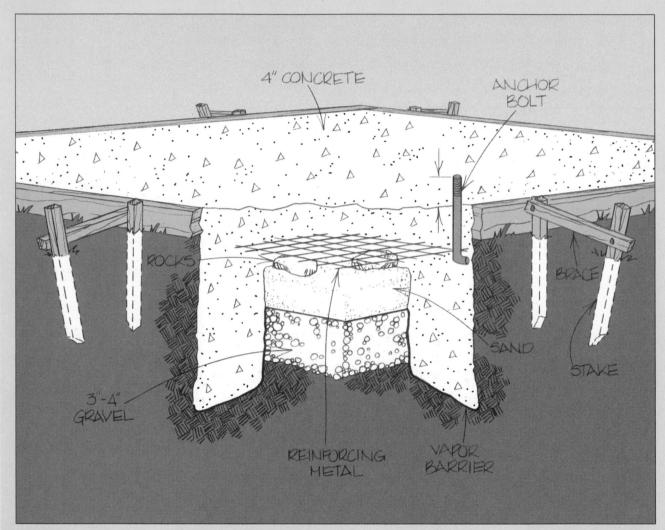

Although a concrete pad looks like a simple slab of stone, it's actually a multilayered structure. It sits on top of layers of sand and gravel. A footer extends around the perimeter to give the pad strength and stability. A vapor barrier keeps it dry, and steel mesh embedded in the pad reinforces it and keeps it from cracking. The thickness of the pad and the width and depth of the footers depend on how the pad is used.

Use	Examples	Pad	Footer	
Light	Garden shed, playhouse, greenhouse	4″	8″ wd. x	12″–16″ dp.
Medium	Patio, utility barn, gazebo	4″	8″ wd. x	18″–24″ dp.
Heavy	Garage, shop	6″	12″ wd. x	24″–36″ dp.

Note: *Never make a pad less than 4″ thick. It won't be strong enough to support its own weight and will crack.*

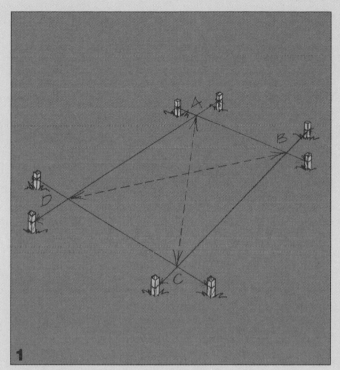

1 Lay out the concrete pad with stakes and string. Locate the corners of the pad where the strings cross. Measure diagonally from corner to corner to check that the layout is square. The distance AC must equal BD.

2 Dig a footer trench around the perimeter of the pad. Make the trench as wide and as deep as you want the footer — the walls of the trench will serve as the form for the footer.

3 Make the form for the pad itself with 2-by lumber. Use 2 x 8s for a 4"-thick pad, and 2 x 10s or 2 x 12s for a 6"-thick pad. Put these in place around the footer trench, and check that the form is square. Level the form by shimming up the corners.

4 Hold the form in place with stakes every 18"–24". These are extremely important! If the form isn't adequately staked, the weight of the wet cement will cause the form to bow out. The resulting pad won't be square.

(Continued)

Step-by-Step: Pouring a Concrete Pad — Continued

5

Remove the grass, plants, and several inches of topsoil from the ground inside the form. Spread a shallow layer of gravel over the bare ground, and a layer of sand over that. Allow room for a pad of the desired thickness. To measure, stretch a string over the top edges of the form and check the distance from the string to the sand at several points. The distance should be the same as or slightly more than the thickness of the pad you intend to pour. If it's less, remove some sand.

6

If you're concerned about dampness, cover the sand with a plastic vapor barrier. (This is not necessary, but it's advisable — especially if you plan to build an enclosed building on the pad.) To keep the concrete from cracking, lay steel reinforcing mesh over the vapor barrier. Raise the mesh 1″–2″ off the barrier with small stones, to allow the cement to flow around and under the wires. (Some concrete companies offer to mix reinforcing strands of fiberglass right in the concrete. If this is available, you won't need mesh.)

9

Continue to thrust the 1 x 2 into the footers for 15–20 minutes, until the concrete just begins to harden. Then have someone help you scrape the excess concrete off the top of the pad; rest a long 2 x 4 on the top edges of the form and move it back and forth as you drag it slowly across the pad. Do this several times.

10

Wait until the concrete hardens a little more — about one hour. Then set ½″ x 10″ anchor bolts along the perimeter wherever you need them. Leave about 2½″ of the **bolt end** protruding from the pad. The concrete should be stiff enough to hold the bolt upright without it tipping over or sinking.

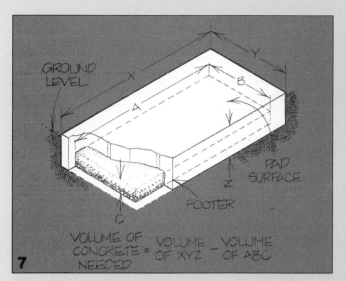

7

Call a local concrete company and arrange to have concrete delivered. (Look in the yellow pages under "Concrete — Ready Mixed.") To figure the amount of concrete you should order, simply calculate the number of cubic yards in the pad, and add 10%–20% to be on the safe side.

8

When the concrete is delivered, spread it out evenly across the pad. Constantly plunge a long 1 x 2 or crowbar into the concrete all around the perimeter to make sure that the footer trenches fill up with no air bubbles. (You can also rent a concrete vibrator to do this.) Have the driver dump the excess into a wheelbarrow — you may need it.

11

Let the concrete harden further — about 1½–2 hours — until it's stiff to the touch. Wet the surface with a garden hose or sprinkling can to make it easier to work. Starting at the center and moving toward the edges, smooth the concrete with a trowel or darby. If the pad is so large that you can't reach the center from the edge, place some large scrap boards on the concrete and kneel on them. The concrete should be stiff enough to support your weight on the boards. If it isn't, you need to wait a little longer.

12

Let the pad harden for at least a day. From time to time, wet it down with a garden hose — this will make it harder. Then remove the stakes and the form, and fill in around the pad with topsoil.

Fences and Gates

A fence may serve many practical purposes: It can define your property line, keep dogs out of your garden, make your yard more private and secure, provide support for climbing plants, and on and on. But just as important, it has an aesthetic purpose: The proper fence *design* provides the right setting for your home or garden, just as the right picture frame enhances a painting.

Fences can take many different forms, from simple strands of wire stretched between posts to shaped boards that follow the contour of the land. The type of fence you need for your home will depend on your architectural tastes, what you need the fence to do, and the local building codes. This chapter shows how to build two of the most popular types — a *picket* fence and a *privacy* fence. The picket fence is mostly used to mark a boundary line, partition a yard, confine animals and children, and provide a pleasant background for a flower bed. The privacy fence will do all that, plus hide a yard from prying eyes and provide a windbreak. The design of either fence can be altered simply by changing the shape of the fence boards.

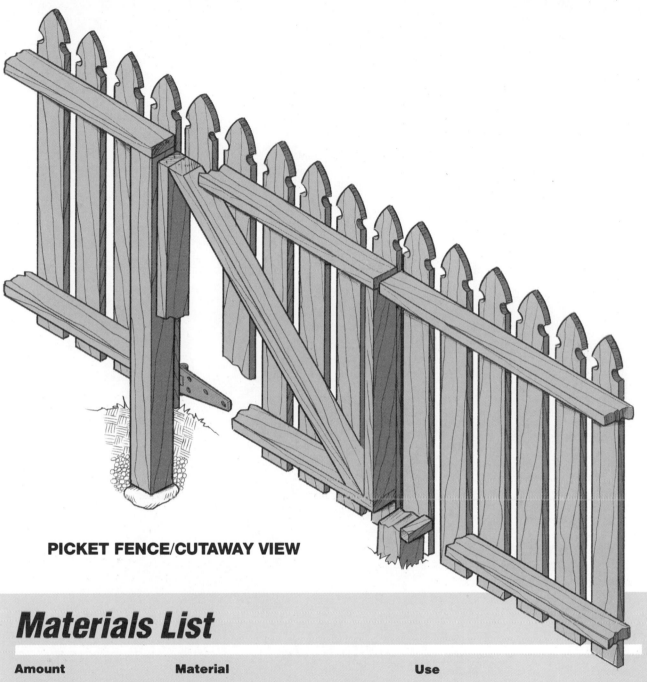

PICKET FENCE/CUTAWAY VIEW

Materials List

Amount	Material	Use
For a 6' section of picket fence:		
1	4 x 4 x 6' Treated LP-22 (cut a 12' post in half)	Fence post
1	2 x 4 x 12' Treated LP-2	Rails
2	1 x 2 x 3½"	Cleats
12–16*	⅝ x 4 x 48" (Cedar or treated LP-2)	Pickets

HARDWARE

8–12	16d Spiral nails (galvanized)	Attach rails
52–68*	8d Spiral nails (galvanized)	Attach cleats and pickets

(Continued)

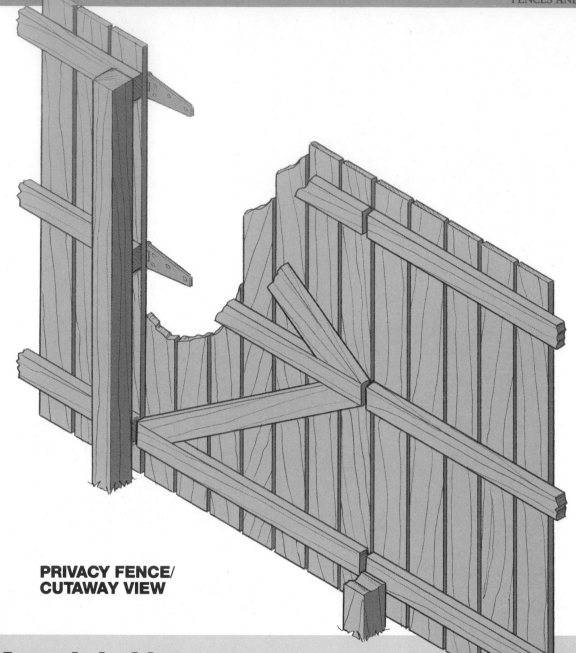

**PRIVACY FENCE/
CUTAWAY VIEW**

Materials List — *Continued*

Amount	Material	Use
For an 8' section of privacy fence:		
1	4 x 4 x 8' Treated LP-22	Fence post
3	2 x 4 x 8' Treated LP-2	Rails
15-16*	⅝ x 5½ x 72" Cedar or treated LP-2	Fence boards

HARDWARE

12–16	16d Spiral nails (galvanized)	Attach rails
90–96*	8d Spiral nails (galvanized)	Attach fence boards

These amounts will vary according to the spacing of the pickets or fence boards.

1 ***Choose a design for your fence.*** All fence designs are constructed in approximately the same manner. The only major difference between the two types shown — privacy and picket fences — is in the way the rails are attached to the posts. On a privacy fence, the rails are nailed directly to the outside of the posts, edge up. On a picket fence, the top rails are nailed to the tops of the posts, and the bottom rails rest on cleats *between* the posts. Both rails are laid flat, faceup (for appearance — the fence looks better this way). Because the rails are faceup on a picket fence, they are more likely to sag. (A board will support more weight when it's edge-up than when it's faceup.) As a result, the span between posts is shorter — a maximum of 6′, compared to 8′ for a privacy fence. There are some minor differences, too. Because a privacy fence

is built to provide seclusion and security, it's usually taller and the boards are spaced closer together.

The way a fence follows the contour of the land often makes a bigger difference in construction than its design. There are three basic ways to build a fence in relation to the surface of the ground. You can make a *straight-top fence,* a *contoured fence,* or a *stepped fence.*

Most fences are either straight-top or contoured. If you have a fairly level piece of property, you'll probably build a fence in which the tops of the boards are level, while the bottoms follow the contour of the land, as shown in the *Straight-Top Fence Detail.* If your land is fairly hilly, it's easier to build the fence so both the tops and the bottoms of the boards follow the land, as shown in the *Contoured Fence Detail.*

STRAIGHT-TOP FENCE DETAIL

CONTOURED FENCE DETAIL

In some instances, you may want to build a stepped fence. Between any two posts — in any one *fence* section — the tops of the boards will all be at the same height. However, the heights will vary from section to section according to the landscape, as shown in the *Stepped Fence Detail*. It's often necessary to build in steps when you're erecting a fence in preassembled sections.

Choose a design for your fence, and decide how it will follow the lay of the land. To build the fence, follow the general procedure outlined in this chapter, and any specific instructions for variations.

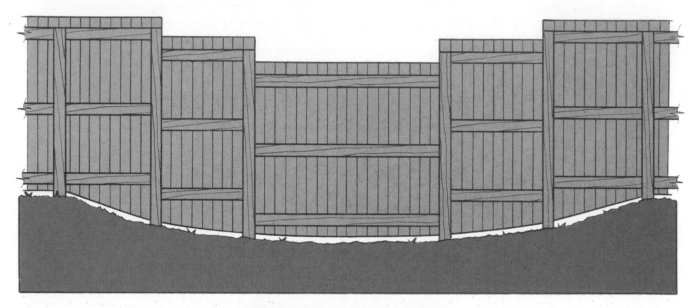

STEPPED FENCE DETAIL

2 **Lay out the fence posts.** Decide where you want to put the fence. If part or all of it will run along your boundary line, locate your survey markers. These are usually permanent metal stakes driven deep in the ground at the corners of your land. If you can't find them, call your local building and planning agency for assistance. Also check the local codes. These will specify how far back you must set a fence from your boundary. Finally, check with the utility and phone companies about any buried gas, power, or phone lines that may cross your land. Some of these may be only 24″ below the surface — about the same depth that you will dig the postholes.

When you've decided exactly where to put the fence, lay out the location of the posts with stakes and string. (Refer to "Step-by-Step: Setting Pole Foundations," page 45.) Space the posts every 8′ or less for a privacy fence, and every 6′ or less for a picket fence. Wherever you want to put a gate, space the posts 30″–36″ apart. For most fence rows, you can simply stretch a string between two stakes and measure along it. However, if your design calls for either precisely square corners or a curved fence, you may have to do a little extra figuring.

Plotting a right angle — Set all the stakes as accurately as you can by eye and stretch string between them. Measure out exactly 3′ from a corner stake along one string, and 4′ along the other. Mark the strings with tape or ribbon, then measure the diagonal between the marks. (See Figure 1.) The diagonal should be exactly 5′. (Remember the Pythagorean theorem from high school?) If it's more or less, move the stakes.

Plotting a curve — Make a giant string compass by tying each end of a long rope to a stake. Decide the radius of the curve, and drive one stake into the ground at the axis. If necessary, adjust the length of the rope by wrapping it around the other stake. Pull the rope tight and, using the tip of the free stake as a pointer, mark the location of the first post in the curved portion of the fence. Swing the compass through its arc several degrees and mark the next. (See Figure 2.) Repeat until you have located all the posts.

You may also have to do a little extra figuring if you're building a contoured or stepped fence on sloping ground.

Spacing posts on hills — First, find the rise and run of the hill — every slope has a rise and a run, just like a stair. The rise is the height of the hill, as measured from the bottom to the crest. The run is the horizontal distance between the crest and the bottom. Mark the highest point of your fence line with a stake. You can eyeball this location; it won't matter if you're off a few inches. Tie a string around the stake, close to the ground. Next, mark the low point of the fence line with a stake that is tall enough to extend above the base of the top stake. Draw the string taut, use a string level to

position it perfectly horizontal, and tie it to the lower stake. (See Figure 3.) Measure the distance from the ground to the string on the lower stake — this is the rise. The distance between both stakes, measured along the string, is the run.

Decide how high you want each fence section to rise. This shouldn't vary much, or the fence will look ridiculous. For fences 4' tall or less, keep the rise under 12". For fences between 4' and 6' tall, keep it under 18".

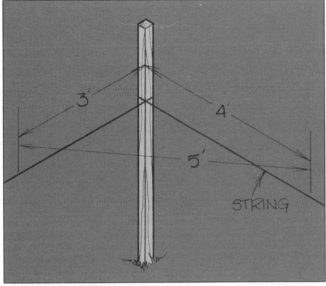

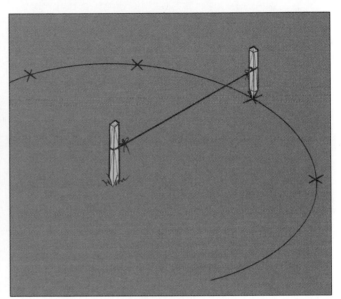

1/Use some simple geometry to lay out a square corner. Mark the strings 3' and 4' out from the corner stake, and measure between the marks. If the measurement is **less** than 5', move the stakes to make the corner angle **larger.** If it's **more** than 5', make the angle **smaller.**

2/Lay out the posts in a curved fence with a giant string compass. To make a graceful curve, space the posts closer together than on the straight portions of the fence.

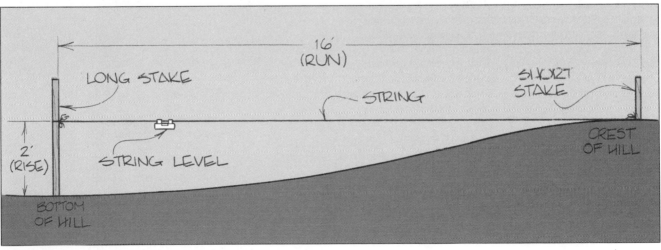

3/Use stakes, string, and a string level to find the rise and the run of a slope. Here, the rise is 2' and the run is 16'.

Divide the desired rise per section into the total rise. For example, if you want each section to rise 6″ and the total rise is 2′ (24″), divide 6 into 24. The answer, 4, is the number of fence sections you'll need to build between the bottom and the crest of the hill. Divide the number of sections into the run to get the length of each section and the spacing between the posts. For example, 4 sections divided into a 16′ run is 4′. Space the posts 4′ apart on center on the hill. (See Figure 4.) Measure along the string and locate the posts with a plumb bob. (Refer to "Step-by-Step: Setting Pole Foundations," page 45.)

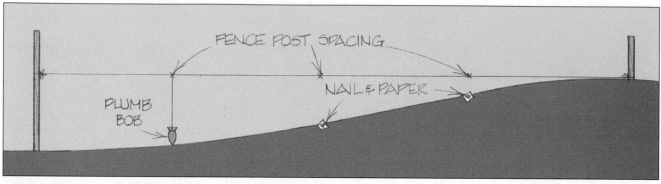

4/After figuring the length of each fence section, measure **along the string** (not along the hill) to space the posts. Transfer the post position to the ground with a plumb bob. Use 16d nails and pieces of paper to mark the post positions.

3 **Set the posts.** When you have located all the posts, dig postholes 6″–8″ in diameter and approximately 24″ deep. Set the posts plumb and fill in the holes. If you're making a straight-top fence, mark the posts to cut them off level with each other. (Refer to "Step-by-Step: Setting Pole Foundations," page 45.)

If you're making a contoured fence on sloping ground, always work your way *uphill* — it's easier. First mark the top of the lowest post, measuring from the ground up. Using a string and a string level, transfer the height of the first to the next post up the hill and add the rise per section that you figured earlier. (See Figure 5.) For example, if the rise per section is 6″, each successive post should be marked 6″ above the previous one. Continue until you reach the post at the crest of the hill.

For a stepped fence, work your way *downhill*. Mark the top of the post at the crest of the hill, then mark the top of the next post down the hill at the *same level*. The drop starts with the third post, and continues one post *beyond* the post at the bottom of the hill. (See Figure 6.)

Once the tops of all the posts are marked, cut them off with a circular saw or hand saw.

Note: If the fence is to be attached to your home or outbuilding, don't place a post right next to the building. Instead, attach a 2 x 4 to the building with ⅜″ x 5″ lag screws, and nail the rails to the 2 x 4 as shown in the *Fence-to-Building Construction Detail*. If the building is made of concrete block or masonry, install expandable lead anchors in the wall to hold the lag screws.

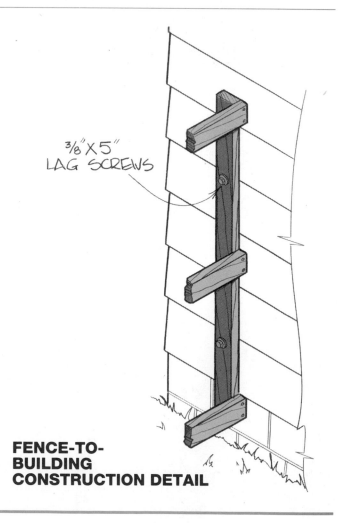

⅜″ X 5″ LAG SCREWS

FENCE-TO-BUILDING CONSTRUCTION DETAIL

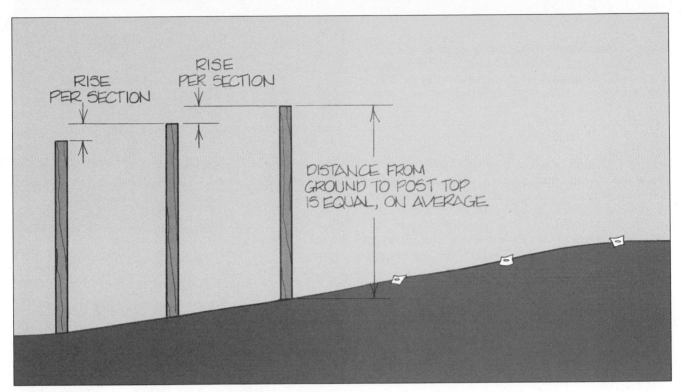

5/On a contoured fence, the tops of all the posts should be approximately the same distance above the ground.

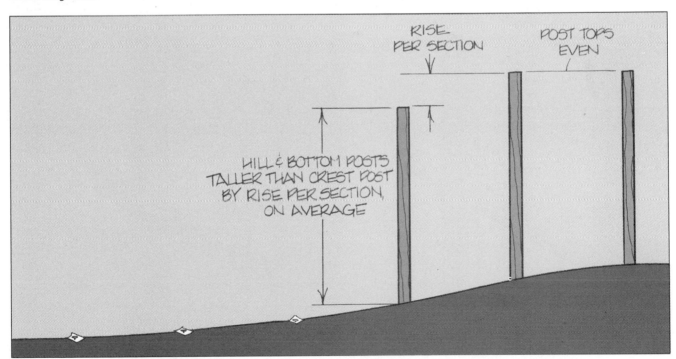

6/On a stepped fence, the posts at the bottom of the hill and the posts on the slope should be **slightly higher** above the ground than the other posts in the fence line.

4

Attach the rails to the posts. Cut the
rails from 2 x 4 stock and attach them to the
posts with 16d spiral nails. If you're making a 4'-high
picket fence, you'll need two rails per section; if you're
making a 6'-high privacy fence, you'll need three.

For picket fences, attach the top rails to the tops of the
posts, as shown in the *Picket Fence Frame Construction* drawing. Measure and mark the locations of the
bottom rails on the posts — they should be 36"–40"
below the top rails, and at least 6" from the ground. Just
below these marks, attach cleats to the facing sides of
the posts, using 8d spiral nails. Rest the bottom rails on
the cleats, and nail them to the posts. (See Figure 7.)

For privacy fences, nail the top rails to the outside of
the posts, flush with the tops, as shown in the *Privacy
Fence Frame Construction* drawing. Measure and mark
the locations of the middle and the bottom rails, then
nail them to the posts. The rails should be spaced
approximately 30" apart, and the bottom rail should
be at least 6" from the ground.

7/The bottom rail
on a picket fence
must be supported
by cleats. However,
nail the cleats **and**
the rails to the
posts. Don't try to
nail the rails to the
cleats — the big
16d nails will
split them.

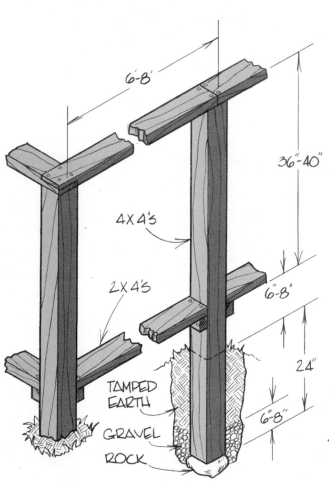

**PICKET FENCE
FRAME CONSTRUCTION**

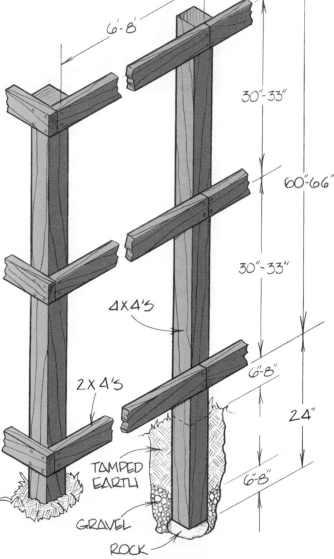

**PRIVACY FENCE
FRAME CONSTRUCTION**

For contoured fences on sloping ground, drop the position of the rails in each section, beginning with the first section downhill from the crest post. For stepped fences on sloping ground, start to drop the rails with the *second* section past the crest post. (See Figures 8 and 9.) With each section, the rails should drop the same distance as the rise per section that you figured.

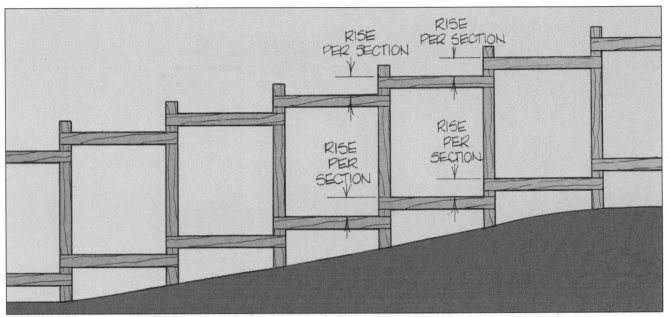

8/When positioning the rails for a contoured fence, drop them slightly with each section past the crest post.

Don't butt the rails together in the center of the posts, as you did on level ground. Cut them to span the width of the posts. This helps to strengthen the fence sections on sloping ground.

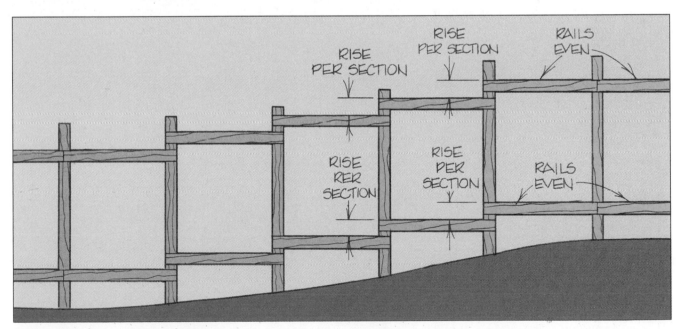

9/When positioning the rails for a stepped fence, don't begin to drop them until the **second** section downhill from the crest post.

5 ***Cut the shape of the pickets or fence boards.*** You can cut any number of different shapes on the top end of each picket or fence board, adding visual interest to the fence. Shown are several possible board top designs for both picket and privacy fences. Some of these are available ready-made from lumberyards. Some can be special-ordered from fence supply companies (look in the yellow pages of the telephone book under "Fence"). You may have to make others yourself.

If you cut your own designs, there are two tricks you can employ to make this chore easier and less time-con-

suming. First, try stacking the fence boards and cutting several at once. This works especially well if you build a fence from cedar or redwood. Both of these are very soft woods, and you can cut a stack of five or six 1-by boards without overloading a power saw. Second, make yourself templates or jigs to help cut identical shapes without having to measure and mark for each cut. Figures 10 through 13 show how to set up a drill press and a radial arm saw to mass produce Gothic fence boards. The necessary jigs are very simple — all you need are fences and stops.

  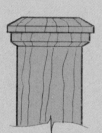 

GOTHIC   ROUNDED POINT   SPADE   POINTED   DOG EAR   POINTED   ROUNDED   SAWTOOTH

PICKET FENCE BOARD TOP DESIGNS   **PRIVACY FENCE BOARD TOP DESIGNS**

TRY THIS! While you're cutting shapes in the tops of fence boards, you may want to make matching or complementary shapes for the post tops. All of these designs can be built up from mold-

ings, turnings, or band-sawed parts. Attach them to the tops of the posts with spiral nails or double-pointed lag screws.

FOR PRIVACY FENCE

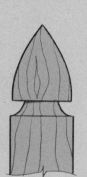  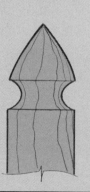 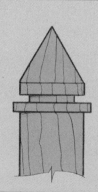

FOR PICKET FENCE

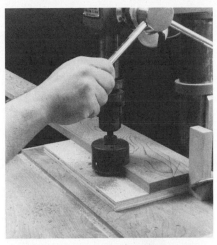

*10/Mount a 2"
circle saw in a drill
press, clamp a
fence to the table,
and clamp a stop to
the fence. Position
the fence and stop
to hold the board
so the saw will cut
a **half-circle** in one
edge, 4" from the
end. Cut a half-
circle in one edge...*

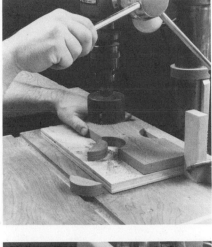

*11/...then turn the
board over and cut
an identical half-
circle in the op-
posite edge.*

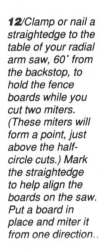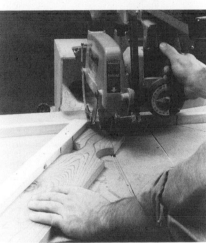

*12/Clamp or nail a
straightedge to the
table of your radial
arm saw, 60° from
the backstop, to
hold the fence
boards while you
cut two miters.
(These miters will
form a point, just
above the half-
circle cuts.) Mark
the straightedge
to help align the
boards on the saw.
Put a board in
place and miter it
from one direction..*

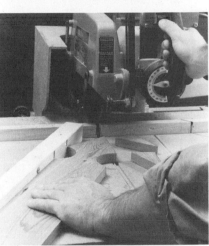

*13/...then flip it
over and miter it
from the other direc-
tion. The miter cuts
will form a point at
the top end of the
board.*

6 Nail the pickets or fence boards to the rails.

For a straight-top fence, attach a picket
or fence board at both ends of a fence row, using 8d
spiral nails. Cut the pickets or boards to the proper
length as you install them. Be careful to cut the boards
so the bottom ends are precisely the same distance
from the ground (no less than 2"), and the top ends are
precisely the same distance above the top rails (4"–12").
Stretch a string between the top ends.

Attach more pickets or fence boards between the first
two boards. Measure each board before you put it up;
cut the *bottom* end so that the top will be even with the
string and the bottom will be a uniform distance from
the ground. (See Figure 14.)

As you work, space the boards evenly. Pickets should
be 1½"–3" apart, and privacy fence boards should be
¼"–½" apart. Cut a scrap of wood of that size and use it
as a gauge to help space the fence boards consistently.
As you approach the end of a fence row, adjust the
spacing slightly (if necessary) so you don't have to split
a board or bunch them up in one place.

*14/Nail up the
fence boards so
the tops are even
with the guide
string. This string
must be stretched
tight enough so*

*that it doesn't sag.
If it's too loose, the
completed fence
will appear to dip in
the middle. If you
can't stretch the
string tight enough*

*to keep it from sag-
ging, divide the
fence row into
smaller portions
and do one portion
at a time.*

TRY THIS! Nailing boards to a long rail can be a trial — the rail bounces with each hammer blow. To prevent this, back up the rail with a "nailing anvil." This is a chunk of steel, sized to fit in your hand, with at least one flat side. (An old sledge hammer head works nicely.) The mass of the steel absorbs the hammer blow and keeps the rail from bouncing. Nailing anvils are also handy for straightening and clinching nails.

For stepped fences, follow this same procedure *for each section*. Run another guide string every time there is a drop or a rise. For contoured fences, cut the fence boards all the same length, and attach them to the fence so the bottoms *and* the tops are all approximately the same distance from the ground.

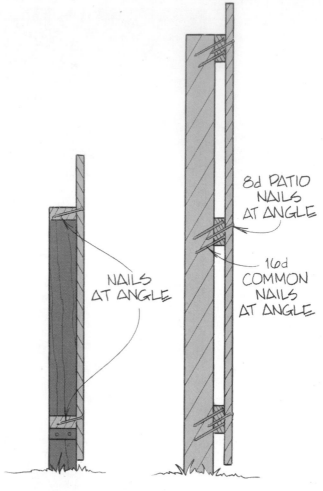

NAILS AT ANGLE

8d PATIO NAILS AT ANGLE

16d COMMON NAILS AT ANGLE

PICKET FENCE CONSTRUCTION DETAIL

PRIVACY FENCE CONSTRUCTION DETAIL

7 Install gates, where needed. The gates for picket and privacy fences are similar. Plan each one to allow ¼″– space on either side, between the gate and the gate post. Arrange the battens so they are even with the fence rails. Then nail the fence boards to the battens, and brace the gate to hold it square. Select hardware that helps to keep the gate from sagging. Look for long, strong strap hinges and a bracket-type latch that secures *and* supports the gate when it's closed. Hinge the gate between the gate posts, and install the latch.

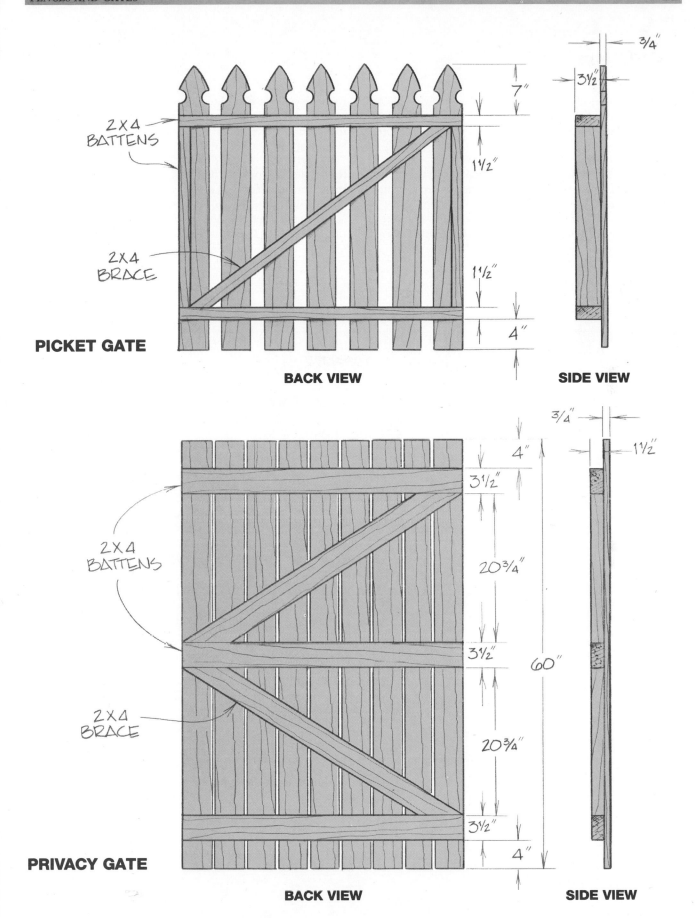

2X4
BATTENS

2X4
BRACE

PICKET GATE

7"

1½"

1½"

4"

3½"

¾"

BACK VIEW   SIDE VIEW

2X4
BATTENS

2X4
BRACE

PRIVACY GATE

4"

3½"

20¾"

3½"

60"

20¾"

3½"

4"

¾"

1½"

BACK VIEW   SIDE VIEW

Variations

As mentioned earlier, picket and privacy fences are just two of the most popular types. Of the many other designs, here are three relatively common ones:

Scalloped-top fence — Build this fence in exactly the same manner as a straight-top privacy fence, but place the top rails approximately 12″ below the tops of the posts. After attaching the fence boards, cut the top ends to follow gentle ellipses, as shown.

Alternating-board fence — This is a privacy fence built on an enlarged picket fence frame. Attach the fence boards to alternate sides of the rails, so the edges appear to overlap slightly. The advantage of this type of fence is that it provides privacy without stopping a gentle breeze from blowing through.

Woven fence — You don't need to put up any rails for this fence. Instead, nail long cleats to the posts, then attach the fence boards *horizontally* to the cleats. Insert a spreader between the fence boards, so one bows out, the one below it bows in, and so on. This makes the fence look as if it were woven.

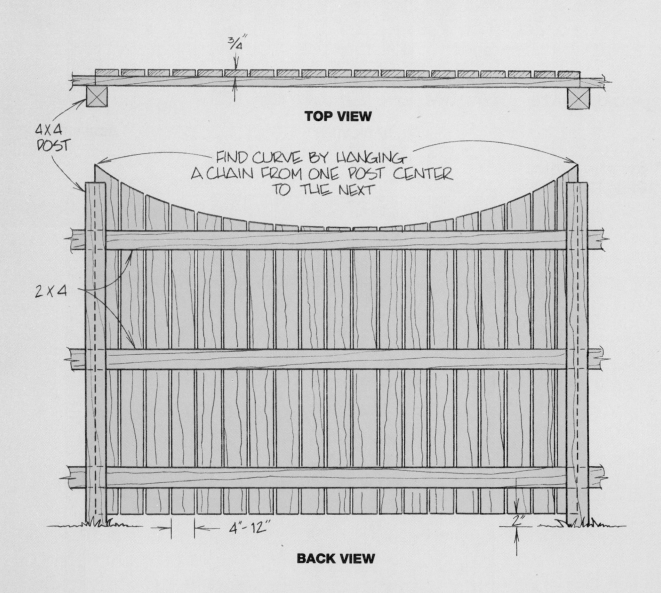

SCALLOPED-TOP FENCE

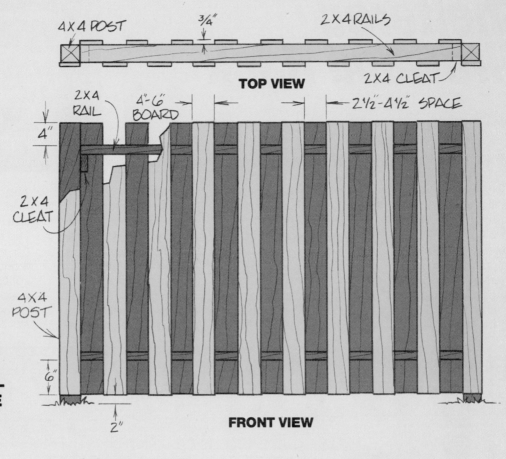

4X4 POST   3/4"   2X4 RAILS

TOP VIEW

2X4 CLEAT

2X4 RAIL   4"-6" BOARD   2½"-4½" SPACE

4"

2X4 CLEAT

4X4 POST

6"

2"

ALTERNATING-BOARD FENCE

FRONT VIEW

4X4 POST   1½" SQ. SPREADER

TOP VIEW

1½" X ¾" CLEAT

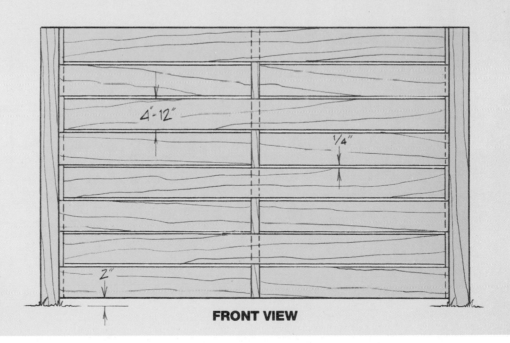

4"-12"

¼"

2"

WOVEN FENCE

FRONT VIEW

Sheltered Firewood Rack

Firewood burns better when it's dry — everyone knows that. So stack it under cover, out of the rain and snow. However, if you put the wood in your garage or under the house eaves, termites and other insects that hide under the bark may find their way into your home. Furthermore, firewood creates a mess — the bark and the sawdust fall off every time you remove a log from the stack. So it's best to store firewood *away* from the house.

This sheltered rack is an inexpensive way to keep firewood neat and dry. It's just a simple frame, supported by four posts and covered by a small roof. However, the rack will hold more than a half cord of firewood, storing it at a comfortable height so even the lowest logs are within easy reach.

The posts keep the stack from tumbling down when you remove a log, and the roof keeps most of the weather off the wood. If any water does collect on the wood, the raised frame lets it drain off quickly.

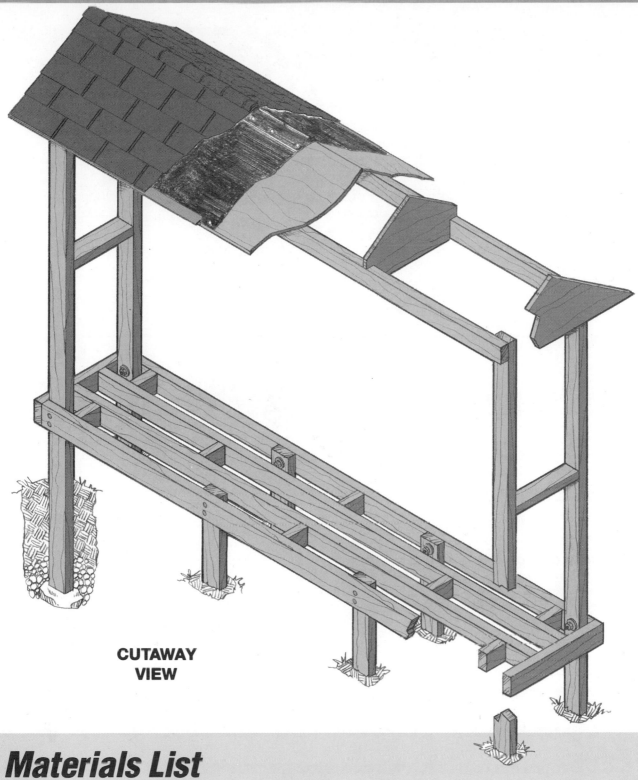

CUTAWAY
VIEW

Materials List

Amount	Material	Use
6	2 x 4 x 10′ Treated LP-22	Supporting posts
7	2 x 4 x 8′ Treated LP-2	Bottom frame, roof frame, braces
2	¾″ x 4′ x 8′ CDX plywood	Trusses, roof sheathing

(Continued)

Materials List — Continued

Amount	Material	Use
HARDWARE		
2 lbs.	16d Common nails (galvanized)	Assemble frames and braces
1 lb.	¾" Roofing nails	Attach roofing materials
32	#10 x 1¼" Flathead wood screws	Attach end trusses and sheathing
8	#10 x 2½" Flathead wood screws	Attach middle trusses
30 sq. ft. (1/6 roll)	Roofing paper	Roofing
2	10' Drip-edge strips	Roofing
66 sq. ft.(2 bundles)	Asphalt tab shingles	Roofing

Note: *This list was figured for the project **as shown in the drawings**. If you change the size of the rack, you may have to adjust the amounts of materials.*

1 Decide how large to make the rack.

As shown, this rack will hold approximately ½ cord of firewood. Depending on how much wood you burn each year, you may want to adjust the size up or down. If you adjust it *up*, you'll also need to beef up some of the frame members. To support any more than ½ cord, make the supporting posts from 4 x 4 stock, and the bottom frame from 2 x 6 stock.

When you have decided how large to build the rack, adjust the dimensions on the plans and make any necessary amendments to the Materials List.

2 Build the bottom frame.

Cut the bottom frame members from 2 x 4 stock, as shown on the *Bottom Frame Layout*. Assemble these parts with 16d nails. First, nail the outside spacers to the two middle rails. Second, nail the middle rails to the middle spacers, joining the two assemblies. Third, nail the outside rails to the outside spacers. Finally, nail the end rails to the outside and middle rails.

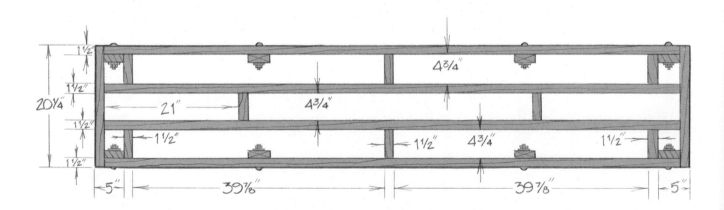

BOTTOM FRAME LAYOUT

3

Set the support posts. Place the floor frame on the ground where you want to build the firewood rack. Make sure it's *away* from other structures. Mark the locations of both the long and the short support posts with stakes. Lift the floor frame over the stakes and set it aside.

Dig postholes, 5"–6" in diameter and 18"–24" deep. Place a large, flat rock or brick in the bottom of each hole and tamp it down. (Refer to "Step-by-Step: Setting Pole Foundations," page 45.) Put the frame back in place — over the holes — and set the posts in the holes, *through* the floor frame.

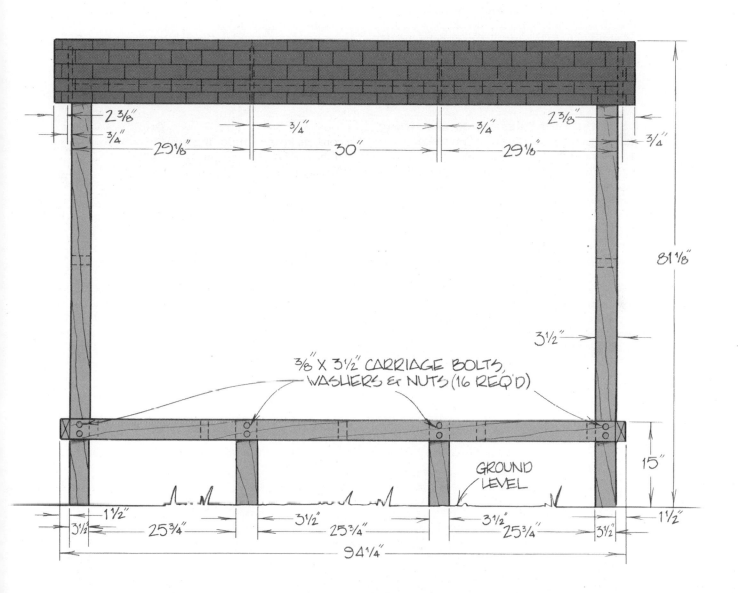

FRONT ELEVATION

4 **Level the bottom frame and attach the posts.** Prop up the frame on concrete blocks, bricks, or wood scraps so it's 11″–12″ above the ground. Using wedges and shims, level the frame as best you can. Straighten the long posts in the holes, brace them so they're plumb, then attach them to the frame with ⅜″ x 3½″ carriage bolts.

Don't attach the four short posts until you cut them to their proper length. Hold the posts straight and mark the tops flush with the frame. Remove the posts, cut them off at the marks, and return them to their holes. Straighten the posts and attach them to the frame with carriage bolts.

5 **Attach the braces and top rails.** Mark each long post 62½″ up from the top edge of the frame, as shown in the *Side Elevation*. Cut all four long posts even with each other, using a circular saw or a hand saw. Carefully measure the distances between the

posts, then cut braces and top rails from 2 x 4 stock. Attach the braces and rails with 16d nails.

Double-check that the four long supporting posts are plumb and that the frame is square. Then fill in the postholes with gravel and dirt, and tamp the dirt down.

6 **Attach the roof trusses and sheathing.** Make the roof "trusses" from ¾″ plywood. These aren't actually trusses, just large triangular pieces of wood, as shown in the *End Truss Layout* and *Middle Truss Layout*. Cut the triangles with a circular saw, then make the notches in the middle trusses with a saber saw or a hand saw.

Attach the end trusses to the long posts with #10 x 1¼″ flathead wood screws, driving the screws through the plywood and into the posts. Attach the middle trusses in a similar manner, but drive #10 x 2½″ screws through the top rails and into the trusses. (See Figure 1.) The bottom edges of all four trusses should be flush with the bottom edges of the top rails, and the gable peaks of the trusses should line up.

1/Attach the trusses and the sheathing with screws. This will hold the parts together much more securely than nails. Ordinary nails, when driven into the **edge** of a plywood part, will pull out easily.

Cut two 8′-long strips of ¾″ sheathing — one strip to cover each side of the roof. Position the sheathing so it overlaps each end truss 2⅜″, then attach it to the trusses with #10 x 1¼″ flathead wood screws.

7 **Apply the roofing materials.** Cover the sheathing with roofing paper, using ¾″ roofing nails to hold the paper down. Install drip edge around

the perimeter of the roof, and cover both the roofing paper and drip-edge strips with tab shingles. (Refer to "Step-by-Step: Installing Tab Shingles," page 101.)

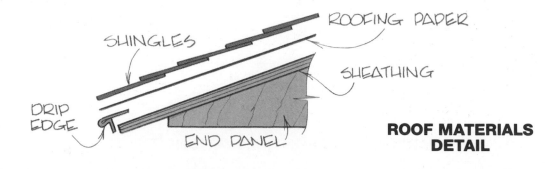

SHINGLES

ROOFING PAPER

SHEATHING

DRIP EDGE

END PANEL

ROOF MATERIALS DETAIL

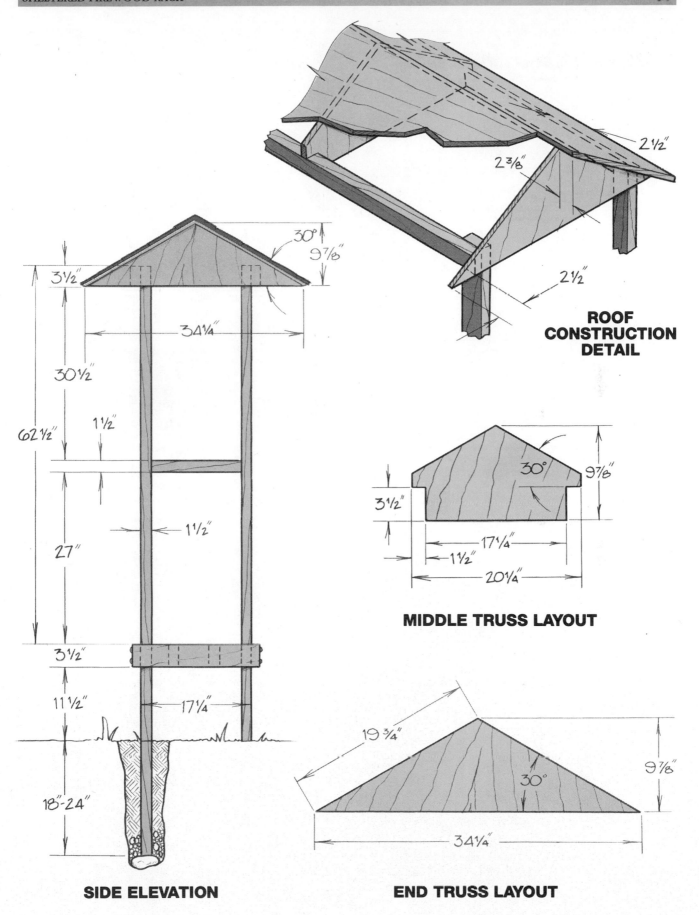

ROOF CONSTRUCTION DETAIL

2½"

2³⁄₈"

2½"

MIDDLE TRUSS LAYOUT

30°

9⁷⁄₈"

3½"

17¼"

1½"

20¼"

END TRUSS LAYOUT

19¾"

30°

9⁷⁄₈"

34¼"

SIDE ELEVATION

30°

9⁷⁄₈"

3½"

34¼"

30½"

1½"

62½"

1½"

27"

3½"

11½"

17¼"

18"-24"

Step-by-Step: Installing Windows and Vents

Windows and vents are complex assemblies. Although you can make your own, it's usually less time consuming (and often less expensive) to pur-chase ready-made, standard-size units at a lumber-yard or building supply center. These units are all in-stalled in the same manner.

1

Apply a thick bead of caulk to the back side of the window or vent trim. Fit the unit into the opening from the outside and push until the trim is flush against the siding.

2

Position the window from the inside. Insert wedges or shims around the unit so it's snug in the opening. Check the assembly with a square. If you're installing a window, also check that it opens and closes easily. If the unit isn't square or the window sticks, remove the appropriate shims and replace them with thicker or thinner pieces.

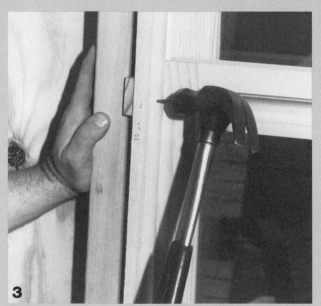

3

When the unit is square, attach the window or vent to the structure by driving several 10d finishing nails through the sides of the assembly and into the wall studs. Leave the wedges and shims in place.

4

From the outside, fill in any gaps between the trim and siding with more caulk.

Step-by-Step: Building Doors and Gates

Most fence gates and outside doors must be custom-built to fit the structure. The easiest way to do this is to build a frame, then cover it with boards, plywood, or siding. However, sometimes the open-ings are not precisely square — this is especially true of gate openings. The following construction method compensates for this imprecision:

1

To make a gate or door, measure the opening and subtract ¼" – ½" from the dimensions. Build a frame from 2 x 4s, 2 x 2s, and 1 x 3s, as necessary, nailing the boards together. Don't cover the frame yet.

2

Wedge the frame in the opening in the fence or building to make sure it fits. There should be a small gap between the frame members and the sides of the opening. Tack braces to the corners to hold the frame rigid, then remove it from the opening.

3

Cover one side of the frame — the side without the braces — with fence boards, plywood, or siding. Then remove the braces. If you're making a door, tack blocks to the inside surface wherever you plan to attach hinges, catches, or other hardware. If needed, install insulation between the frame members. Then cover the other side of the door frame and fasten decorative trim to the outside surface of the door.

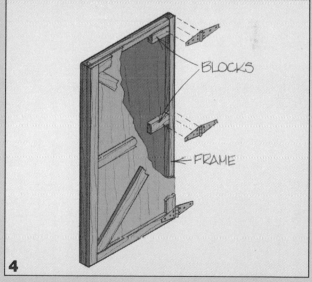

4

BLOCKS

FRAME

Mount the gate or door on hinges. Use T-hinges or strap hinges, since these provide more support than ordinary butt hinges. Position the hinge so the mounting screws will bite through the fence boards or door covering and into the frame members or blocks.

Trash Can Corral

Trash cans have the annoying habit of wandering off or tipping over. When they're empty, a stiff breeze can send them rolling down an alley. When they're full, a dog or raccoon can overturn them and scatter the contents. The lids, too, often leave for parts unknown. When the lids are gone, the cans fill up with rainwater and make garbage soup!

You can avoid this trouble by making a simple enclosure or "corral" to keep your cans and lids in place. This trash can corral is a small wooden platform surrounded by a railing. The platform is just big enough to hold the cans, and the railing just high enough to keep them from tipping over. Part of the railing is hinged so you can easily take the cans in and out of the enclosure. Attach the lids to the frame with eye screws and rope or chain. Then it's almost impossible to lose them.

You can attach a corral to the house or garage, as shown, or you can build a freestanding structure, independent from a building. As drawn, the corral will hold four standard-size (30 gallon) cans, but you can easily adapt the design to hold almost any quantity or size of trash cans.

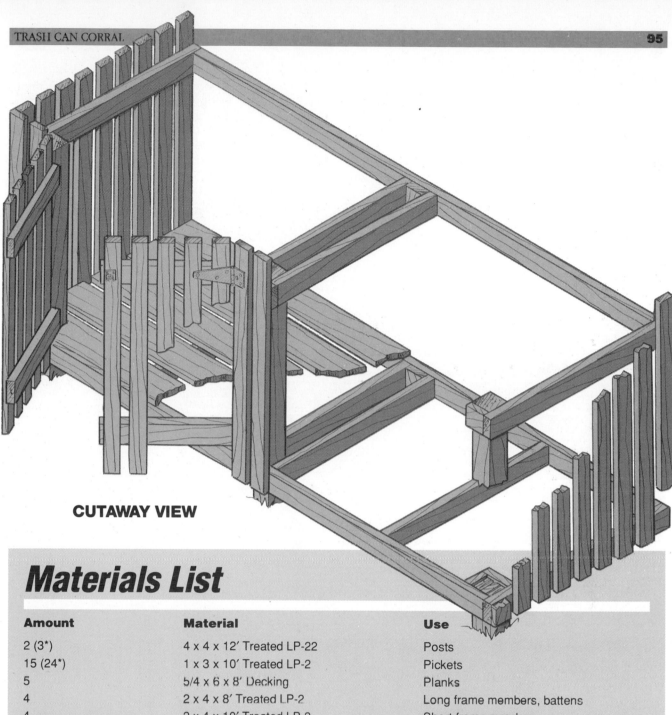

CUTAWAY VIEW

Materials List

Amount	Material	Use
2 (3*)	4 x 4 x 12' Treated LP-22	Posts
15 (24*)	1 x 3 x 10' Treated LP-2	Pickets
5	5/4 x 6 x 8' Decking	Planks
4	2 x 4 x 8' Treated LP-2	Long frame members, battens
4	2 x 4 x 10' Treated LP-2	Short frame members

HARDWARE

Amount	Material	Use
1 lb.	16d Common nails (galvanized)	Assemble frame
3 lbs.	8d Spiral nails (galvanized)	Attach pickets
4 pair	4" T-hinges and mounting screws	Hang gates
2	Hasps and mounting screws	Secure gates
4	1" Eye screws	Anchor ropes or chains
10'	3/8" Rope or light chain	Secure can lids

* Purchase this amount if you build a freestanding unit.

Note: *This list was figured for the project **as shown in the drawings.** If you change the size and configuration of the corral, you may have to adjust the amounts of materials.*

1

Determine the size and configuration of your corral. Decide where you want to put this trash can corral, how many trash cans you want it to hold, and whether it will be attached to a building or freestanding. As designed, the corral will hold four standard-size cans in a single row. To adjust the design to hold larger or smaller cans, more or fewer cans, or hold them in a different arrangement, simply increase or decrease the length and width of the corral. To build a freestanding unit, add posts and pickets to the back side, where shown in the *Top View*.

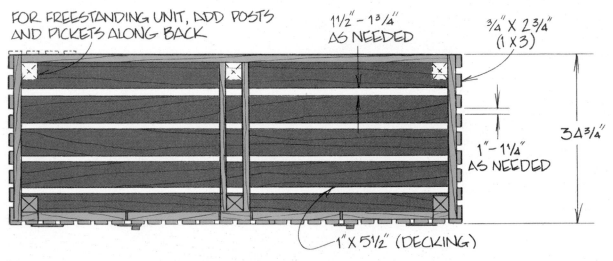

FOR FREESTANDING UNIT, ADD POSTS AND PICKETS ALONG BACK

1½" – 1¾" AS NEEDED

¾" X 2¾" (1 X 3)

34¾"

1" – 1¼" AS NEEDED

1" X 5½" (DECKING)

TOP VIEW

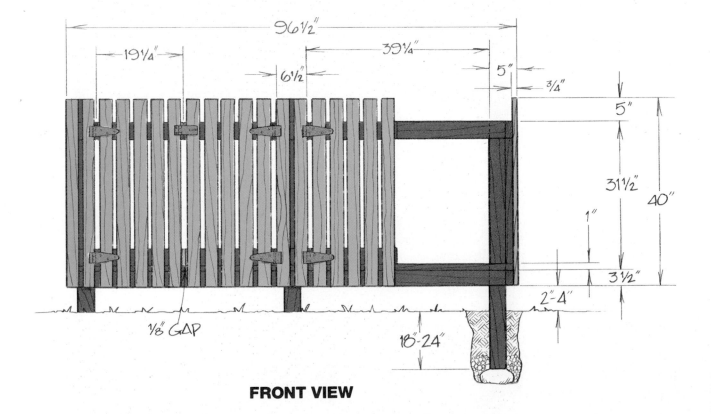

96½"

19¼"

6½"

39¼"

5"

¾"

5"

31½"

40"

1"

3½"

2"– 4"

18"– 24"

⅛" GAP

FRONT VIEW

2

Assemble the bottom frame. From
2 x 4 stock, cut the parts for the bottom frame.

Assemble just the four *outside* frame members — front,
back, and sides — using 16d nails.

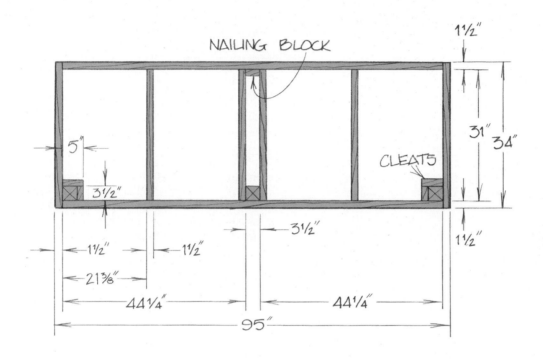

NAILING BLOCK

1½"

31"

34"

5"

3½"

CLEATS

3½"

1½"

1½"

1½"

21⅜"

44¼"

44¼"

95"

**BOTTOM
FRAME
LAYOUT**

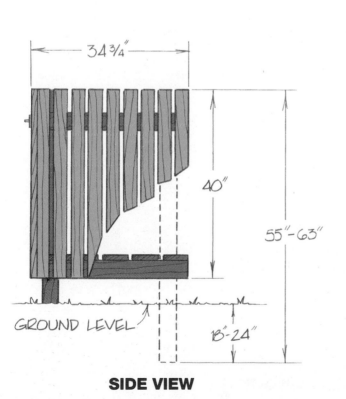

34¾"

40"

55" - 63"

GROUND LEVEL

18" - 24"

SIDE VIEW

3 Set the posts.
Place the frame assembly on the ground where you want to build the corral. Check that the frame is square by measuring diagonally from corner to corner. Make two measurements, from opposite pairs of corners; the frame is square when the measurements are equal.

Using the frame as a guide, mark the location of the posts with stakes or large nails. Remove the frame and dig the postholes 6″–8″ in diameter. Since this is a small structure, the holes need to be only 18″–24″ deep. When you finish digging, replace the frame over the holes.

Cut 6′-long posts from the 4 x 4 stock. Put a large, flat rock in the bottom of each hole, then set a post on top of the rock. All the posts should be inside the frame. *Don't* fill in the postholes yet.

4 Join the frame and the posts.
Prop the frame up on scrap wood or bricks so that it's 2″–4″ off the ground. Check that the frame is level. Should you find that one part is lower than another, raise it with wooden wedges or shims.

If you intend to attach the corral to a building, secure one side of the frame to the wall. To do this, first locate the wall studs, then drive ⅜″ x 5″ lag screws through the frame from the inside and into at least two separate studs. If the wall is made of block or masonry, sink expandable lead anchors in the wall to hold the lag screws.

Lightly clamp the posts to the frame, then brace them plumb in the holes. (Refer to "Step-by-Step: Setting Pole Foundations," page 45.) Each post must stand straight up *and* down and fit snug against the frame. Check the frame once more to be sure that it's still level, and secure it to the posts with 16d nails.

5 Cut the posts to length.
For each post, measure from the top edge of the bottom frame up to where you want to cut it off. (If you're building this corral for standard-size trash cans, the posts should stand 31½″ above the bottom frame.) Mark the post and cut it off with a hand saw or circular saw. Repeat for the other posts, cutting them all to the same height.

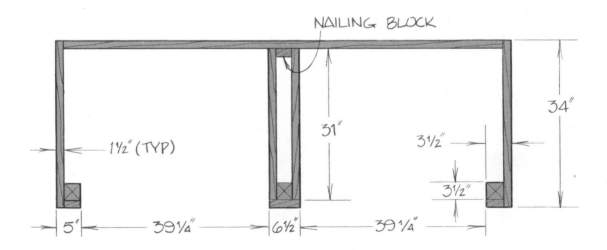

TOP FRAME LAYOUT

6

Assemble the remaining frame members. With 16d nails, assemble the remaining parts of the bottom frame — joists, cleats, and (if needed) nailing blocks. (See Figure 1.) Then cut the top frame members and attach them to the posts. If

you're building an attached corral, secure the one side to the building in the same manner that you did the bottom frame. When the frame is complete, fill in around the posts and tamp the earth down. (See Figure 2.)

1/Nail the cleats to the posts flush with the top edge of the bottom frame. These cleats help to support the platform planks.

*2/Don't fill in the postholes until **after** you've completed the frame. Sometimes, no matter how carefully you set the posts, you must adjust their positions slightly as you build. It's much easier to do this if you haven't filled the postholes yet.*

7

Attach the planks to the bottom frame. Cut the planks from 5/4 x 6 decking, then nail them to the bottom frame with 8d spiral nails. (See Figure 3.) When installed, the planks should be flush with the outside edge of the frame.

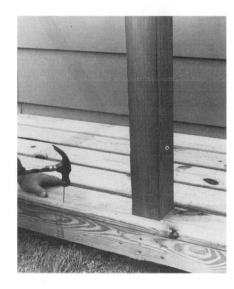

3/Where necessary, notch the planks to fit around the posts.

8

Attach the pickets to the frame. Cut as many pickets as you need from 1 x 3 stock, and decide how far apart to space them on the frame. Depending on the size of the corral, you may find it hard to determine this spacing. However, if you're careful, you can locate them almost as precisely by eye.

Lean the pickets against the top frame and shift them back and forth until they seem to be evenly spaced.

when you're satisfied with the arrangement, mark their locations on the top frame. Attach them with 8d spiral nails, but *don't* drive the nails all the way home yet.

After you've put up all the pickets, take a good look at the corral from all angles. If the spacing seems uneven at any point, pull the nails and move the pickets. When you're satisfied with the way the pickets look, hammer the nails all the way into the wood.

9

Build and attach the gates. Measure the openings in the corral, then cut the gate battens from 2 x 4 stock. Plan to leave a ¼″ gap on each side of each gate. If you're building just one gate for an opening, cut the battens ½″ shorter than the width of that opening. If you're making two, as shown in the drawings, cut them ⅜″ shorter.

Lay out the boards and battens, and assemble them with 8d nails. (Refer to "Step-by-Step: Building Doors and Gates", page 93.)

Although you can add braces to the gates if you wish, they're not needed. The gate is designed so that the bottom batten rests on the platform when the gate is closed. This not only helps keep the gate closed, but also prevents the gate from sagging.

Hang the gates from T-hinges, using mounting screws long enough to go through the pickets and into the battens or frame. Secure the gates with hasps.

10

Tie or chain the lids to the corral. Install eye screws in the top frame, opposite the gates, one per trash can. Cut 2½′ lengths of ⅜″-diameter nylon rope or chain. If you use rope, tie one end to the eye screws and the other to the lid handle; if you use chain, attach it with S-hooks. (See Figure 4.) There should be enough slack in each rope or chain to allow you to put the lid in place when the can is in the corral.

*4/Some plastic trash can lids have solid handles. To secure this sort of lid, cut a small piece of scrap wood and drill a hole through it **and** the lid. Insert the rope through the lid and then the scrap, and tie a knot to keep the rope in place.*

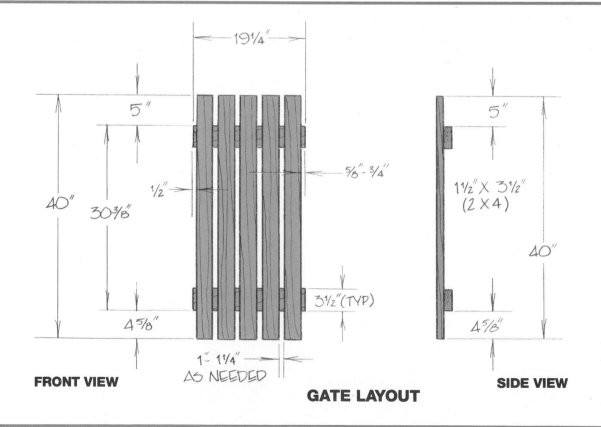

FRONT VIEW   **SIDE VIEW**

GATE LAYOUT

Step-by-Step: Installing Tab Shingles

There are many types of roofing materials, but the most common (and most versatile) are asphalt tab shingles. These usually come with three square-end tabs, as shown, although many different contours are available. They also come in different colors and textures. No matter what the design, they are meant to be installed over plywood sheathing and roofing paper.

Cover the roof frame with plywood sheathing. For most roofs, ½"-thick sheathing is adequate. However, you may want to use thicker material if the roof must weather heavy snowfalls, or if you don't want to see the roofing nails poking through the underside of the sheathing.

Wherever you can, install the plywood sheets so the seams between them are over a rafter. Where you can't do that, put **panel clips** between the sheets to strengthen the roof. Space the clips every 8"–12".

If you want to cover the bottom ends of the rafters (the overhang) with facing, you must do this before you install the drip edge, roofing paper, or shingles. These materials will overlap the facing.

Starting at the bottom edge of the roof and working toward the peak, cover the sheathing with roofing paper or tar paper. Roll the paper out horizontally and tack it down with roofing nails. Overlap each horizontal strip 3"–4". **Note:** You should apply the roofing paper horizontally for two reasons: (1) It won't shed water as well if applied vertically, and (2) it probably is imprinted with lines that will help you align the rows of shingles, but only if the paper is horizontal.

(Continued)

Step-by-Step: Installing Tab Shingles — Continued

5

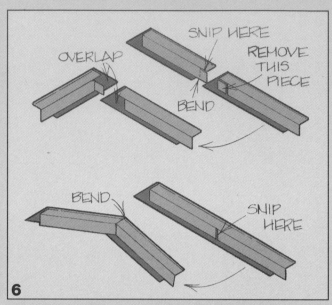

6

Install drip edge all around the perimeter of the roof, nailing the long flange to the sheathing with roofing nails. This T-shaped metal flashing supports the edge of the shingles, allowing them to overhang the roof about ¾″ and protect the wood from rainwater. The flashing also hides and protects the edge of the sheathing (if you haven't installed facing), and helps to hold the roofing paper in place until you apply the shingles.

When turning a corner, snip the drip edge in two and cut interlocking tabs, as shown. When following a roof peak, just snip the short bottom flange and bend the drip edge to fit the roof.

9

10

Apply all subsequent courses of shingles tab side down, overlapping the preceding course by approximately half the width of the shingles. The bottom edge of the upper course should be even with the beginning of the tabs on the course immediately under it. Follow the lines on the roofing paper to make sure each course is straight. Where you can't use the lines, snap a chalk line.

When you reach the peak, fold the shingles on one side over the peak. Then fold the shingles on the other side over those, lapping the two rows of shingles at the peak. Tack the folded shingles down securely with roofing nails.

7

When you're ready to apply the shingles, start at the bottom edge and work your way toward the peak — as you did with the roofing paper. Peel the cellophane strip off the back of each shingle before you apply it. Using roofing nails, lay the first course of shingles tab side up, as shown. (The "tab side" is the side with the slots.)

8

Apply the second course tab side down, directly over the first course. As you work, don't bother to trim the shingles even with the ends of the roof. It's easier to do this after you've applied all the shingles.

11

Cover the peak with a ridge cap made from shingles cut in thirds. (Cut the shingles at every tab-slot.) Decide which end of the peak faces **away** from the prevailing wind. Starting at this end, bend a one-third shingle over the peak so half lays on one side of the roof and half on the other. Nail the shingle in place, then lap another over it, and nail it in place. Repeat, installing a row of overlapping, short shingles along the entire peak.

12

Cut the shingles even with the ends of the roof, using a utility knife or a saber saw with a shingle-cutting blade. This blade looks like a knife — it has no teeth. If you can't find one at your local hardware store, you can make your own by grinding the teeth off a worn-out wood-cutting blade.

Victorian Gazebo

During the last half of the nineteenth century, formal gardens — and garden architecture — became popular among the well-to-do. The Victorian gentry planted acres of flowers, trees, and shrubs and graced them with decorative outbuildings. Perhaps the most popular Victorian garden structure was the "gazing room" — a small, open building from which to survey the garden. Toward the end of the century, an English architect combined *gazing room* with the Latin for "I shall see" — *videbo* — and coined the word *gazebo*.

Gazebos are still popular today, although there's less gazing to do — most contemporary gardens are smaller and less formal than their Victorian counterparts. However, these unique buildings have found other uses. Many serve as detached decks or summer houses, expanding warm-weather living space. They can also be arbors, playhouses, even carports.

The gazebo shown is designed to look like an old Victorian structure, complete with ornamental "carpenter gothic" trim. However, it's built with many of the same methods and materials used in contemporary decks. This simplifies construction and keeps the cost of the project down.

CUTAWAY VIEW

Materials List

Amount	Material	Use
9	4 x 4 x 12′ Treated LP-22	Posts, floor supports, newel posts, key block
4	2 x 10 x 8′	Ornaments
6	2 x 8 x 10′ Treated LP-2	Floor frame
8	2 x 6 x 8′ Treated LP-22	Rafters
14	2 x 4 x 10′ Treated LP-2	Floor joists, top plate, cap plate, stair frame
9	2 x 4 x 8′ Treated LP-2	Railings
37	5/4 x 6 x 10′ Decking	Floor planks, slats, stairs
8	1 x 4 x 10′ Treated LP-2	Fascia, ledgers
8	¾″ x 4′ x 8′ BCX plywood	Roof

(Continued)

Materials List — *Continued*

Amount	Material	Use
HARDWARE		
8	³⁄₈″ x 6″ Lag screws	Attach heart ornaments
44	³⁄₈″ x 4″ Lag screws	Attach frames to posts, ornaments to frames
10 lbs.	16d Common nails (galvanized)	Assemble frames
2 lbs.	10d Spiral nails (galvanized)	Assemble plates, attach railings
10 lbs.	8d Spiral nails (galvanized)	Attach decking
3 lbs.	6d Common nails (galvanized)	Attach ledgers, slats
5 lbs.	³⁄₄″ Roofing nails	Attach roofing materials
3⅓ sq.	Asphalt tab shingles	Roofing
1 roll	Roofing paper	Roofing
4	10′ Drip-edge strips	Roofing
1 gal.	Roofing compound	Seal seam between shingles and key block
4 gal.	Exterior primer	Prime and protect raw wood surfaces
3 gal.	Exterior paint	Aesthetics

1

Build the floor frame. Although most outdoor building projects start with a foundation, it's easiest to start this one with the floor frame. This enables you to use the frame to lay out the pole foundation.

Cut the crossbeams to length, and notch them as shown in the *Crossbeam Layout*. (See Figure 1.) Fit the crossbeams together — the notches form half-lap joints. If these laps bind, place a piece of scrap wood on the joint and drive them together with a mallet. The scrap wood protects the beam.

Saw the outside beams, cutting the ends of four of these pieces square, and mitering the ends of the other four at 45°. Attach the four square-end beams to the ends of the crossbeams with 16d nails, as shown in the *Floor Frame Layout*. Then attach the mitered-end beams, toenailing them to the crossbeams.

1/To make each notch, first cut the sides with a saber saw or a hand saw. Then separate the waste with a chisel.

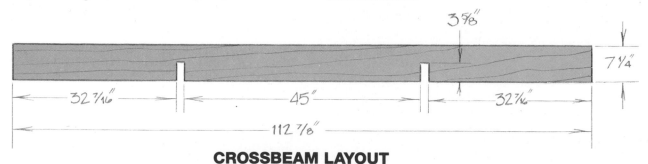

CROSSBEAM LAYOUT

2

Set the posts. With a helper, place the floor frame on the ground where you want to build the gazebo. Determine which way you want the stairs to face and rotate the frame accordingly — you can only attach the stairs to the square-end beams. When you've oriented the floor frame properly, mark the locations of the posts with stakes. Be sure to mark the short floor support posts near the center of the frame. Lift the frame over the stakes and set it aside.

Dig postholes, 6"–8" in diameter and 24"–36" deep. Place a large, flat rock or a brick in the bottom of each hole and tamp it down. (Refer to "Step-by-Step: Setting Pole Foundations," page 45.) Put the frame back in place — over the holes — and set the posts in the holes. (See Figure 2.)

2/Set the posts in the postholes **through** the floor frame. **Don't** straighten the posts or fill in the holes just yet.

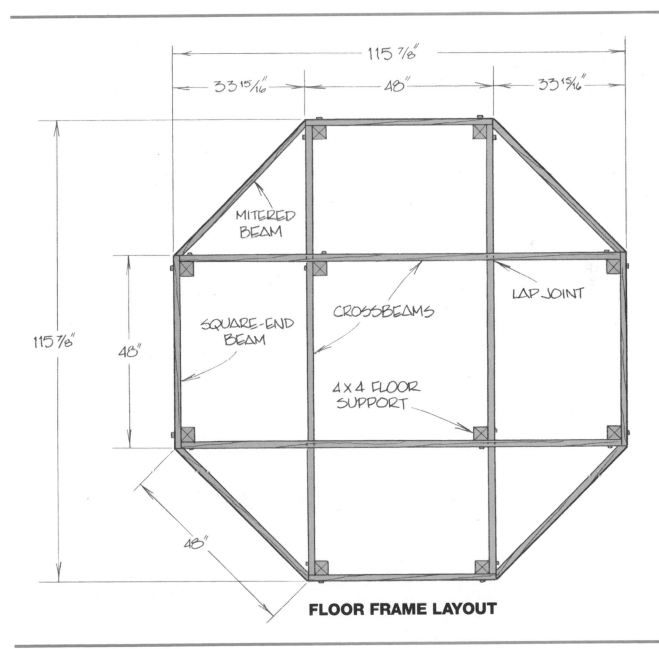

FLOOR FRAME LAYOUT

115 7/8"

33 15/16"   48"   33 15/16"

115 7/8"

48"

48"

MITERED BEAM

LAP JOINT

SQUARE-END BEAM

CROSSBEAMS

4 X 4 FLOOR SUPPORT

3 **Level the floor frame.** Prop up the floor frame at *four* points with concrete blocks, bricks, or wood scraps. It should be about 12″ above the ground. Working with wedges and shims, level it side to side, and then front to back. Double-check side to side (see Figure 3) and repeat until the frame is perfectly horizontal in all directions.

*3/Prop the frame up on blocks, then level it with wooden wedges and shims. Don't attach the posts until **after** the frame is level.*

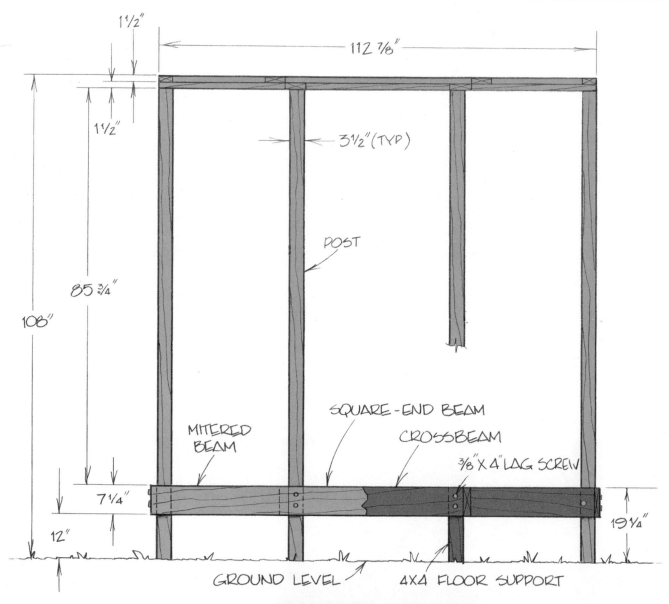

1½″

112⅞″

1½″

3½″(TYP)

85¾″

POST

108″

SQUARE-END BEAM

CROSSBEAM

MITERED BEAM

⅜″ X 4″ LAG SCREW

7¼″

19¼″

12″

GROUND LEVEL   4X4 FLOOR SUPPORT

FRAME/SIDE ELEVATION

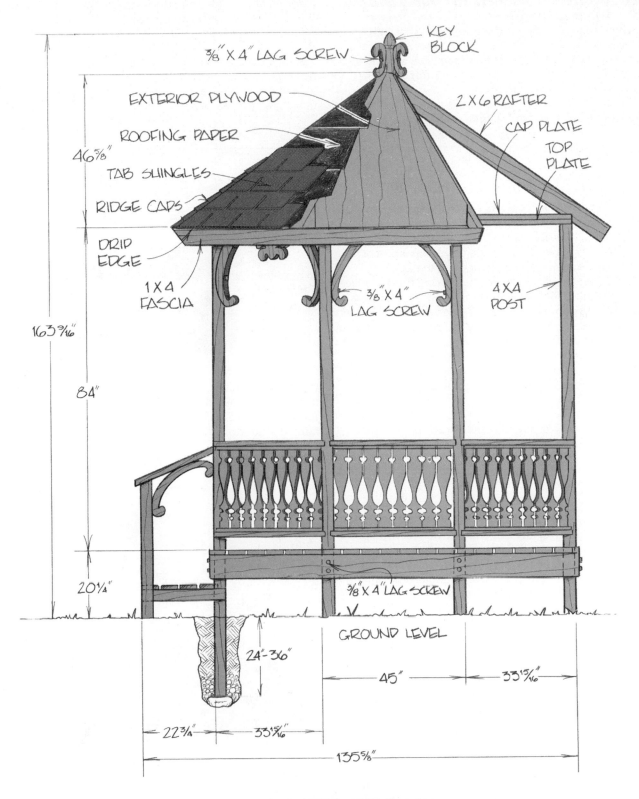

3/8" X 4" LAG SCREW

KEY BLOCK

EXTERIOR PLYWOOD

2 X 6 RAFTER

ROOFING PAPER

CAP PLATE

TOP PLATE

46 5/8"

TAB SHINGLES

RIDGE CAPS

DRIP EDGE

1 X 4 FASCIA

3/8" X 4" LAG SCREW

4 X 4 POST

163 9/16"

84"

20 1/4"

24"-36"

3/8" X 4" LAG SCREW

GROUND LEVEL

45"

33 15/16"

22 3/4"

33 15/16"

135 5/8"

SIDE ELEVATION

4 Attach the posts to the frame and cut them to length.

Straighten the posts so they're plumb, and brace them in place. Attach the frame to the posts with ⅜" x 4" lag screws, driving the screws through the frame from the outside and into the posts. When all the posts are attached, remove the blocks that hold up the frame.

Don't attach the two floor support posts in the interior of the frame until you cut them to their proper length.

Mark the top of the posts flush with the top edge of the frame. Remove the posts, cut them to length, and return them to their holes. Straighten the short posts and attach them to the crossbeams with lag screws.

Mark each long post 85¾" up from the top edge of the floor frame, as shown in the *Frame/Side Elevation*. Cut all eight long posts even with each other, using a circular saw or hand saw.

5 Attach the top plate and cap plate.

From 2 x 4 stock, cut the parts needed for the plates that tie the posts together, as shown in the *Top/Cap Plate Layout*. Saw eight short plates and eight long plates, mitering the ends of each at 45°.

Attach the four *short* plates first, tacking them in place with 16d nails. Don't drive the nails home yet, in case you need to reposition a board. (Just the mitered points of these plates rest on the tops of the posts, as shown in the "Top Plate Only" side of the *Top/Cap Plate Layout*.) When the short plates are all in place, attach the long plates. Once again, don't hammer the nails all the way into the wood until you're sure all the plates are positioned properly.

Reinforce the top plate, driving more nails until there are at least two nails in each end of each board. Attach

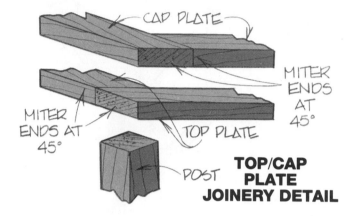

TOP/CAP PLATE JOINERY DETAIL

the cap plate to the top plate with 10d spiral nails, tying the plates together as shown in the *Top/Cap Plate Joinery Detail*. Fill in the postholes with gravel and tamp the dirt down.

6 Assemble and attach the stair frame.

Carefully measure the distance between the posts where you will mount the stair frame. (Remember, you must place the stair between two posts that are spanned by a *square-end* beam.) If necessary, adjust the dimensions on the *Stair Frame/Top Elevation*. Cut the parts for the stair frame from 2 x 4 stock, and assemble them with 16d nails.

Place the stair frame between the posts and mark the position of the newel posts with stakes. Remove the floor frame, dig the postholes, then return the frame to its place. Cut a 4 x 4 x 12' in half to make two 6'-long newel posts. Place the posts in the holes, through the stair frame. Prop the bottom of the frame up about 5⅝" off the ground and level it with wedges and shims. Straighten the newel posts, and brace them so they're plumb. Attach the four corners of the frame to the posts with ⅜" x 4" lag screws. (See Figure 4.)

Double-check that all the posts are plumb and that the frames are level. Fill in the postholes with gravel and dirt, and tamp the dirt down

4/Attach the stair frame to the posts with lag screws, driving the screws from **inside** the frame.

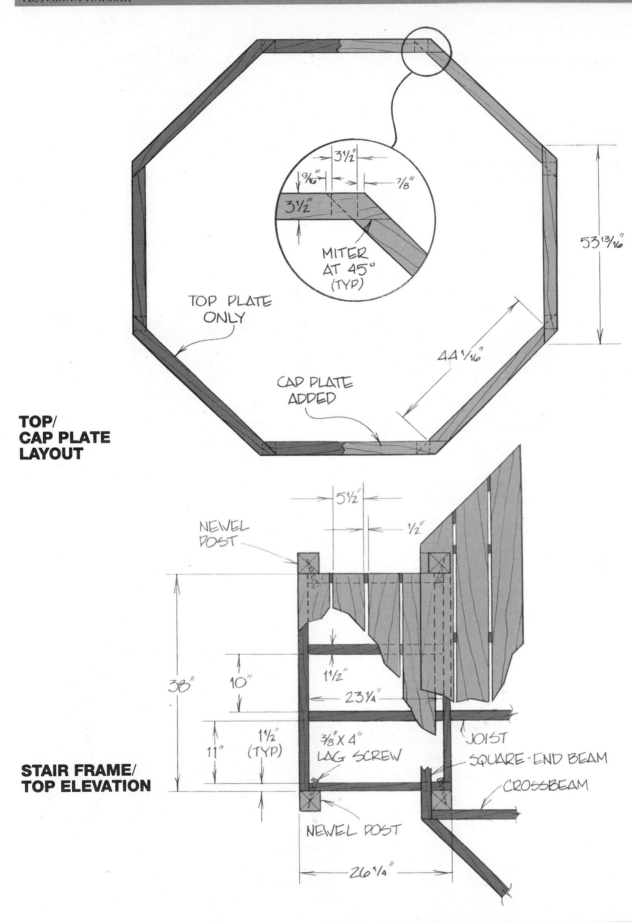

3½"

9/16"

7/8"

3½"

MITER
AT 45°
(TYP)

53¹³/₁₆"

TOP PLATE
ONLY

44¹/₁₆"

CAP PLATE
ADDED

**TOP/
CAP PLATE
LAYOUT**

5½"

½"

NEWEL
POST

10"

1½"

23¼"

38"

11"

1½"
(TYP)

⅜" X 4"
LAG SCREW

JOIST

SQUARE-END BEAM

CROSSBEAM

**STAIR FRAME/
TOP ELEVATION**

NEWEL POST

26¼"

7 Cut and install the floor joists.

Measure and mark the positions of the joists on the floor frame. The joists should run *parallel* to the short dimension of the stair frame, so the planks on both the stairs and the gazebo floor will be parallel. If you put the stairs on the north or south side of the gazebo, run the joists north and south.

Using 1¼″ roofing nails, attach joist hangers to the outside beams and crossbeams. Cut the joists from 2 x 4 stock. The ends are square, except those that join the mitered outside beams; miter these at 45°. Rest the joists in the hangers and secure them with roofing nails or 16d nails, as necessary. (See Figure 5.)

To support the floor planks around the posts, cut cleats from 2 x 4 scraps. Nail them to the posts with 16d nails, as shown in the *Floor Layout* and Figure 6.

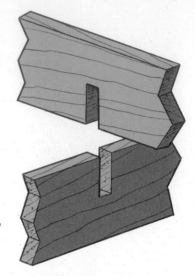

FLOOR JOIST JOINERY DETAIL

5/Support all the **square** ends of the joists with joist hangers. Nail the mitered ends to the beams with 16d nails.

6/Attach 2 x 4 cleats to the posts, flush with the top edge of the floor frame. These will support the flooring around the posts.

8 Install the floor planks.

Using 8d spiral nails, attach the floor planks to the frame, perpendicular to the joists. Space each plank ½″ from those on either side, so all planks have room to expand and contract with changes in the weather. Where necessary, notch the planks to fit around the posts. (See Figure 7.)

7/Don't bother to cut the planks to length as you attach them. Let them hang over the edge. When you've installed them all, cut them off even with the outside of the frame.

9 Assemble and attach the roof frame.

Cut eight rafters as shown in the *Rafter Layout*. Also, make an octagonal key block from a scrap of 4 x 4 stock. Make the four additional faces on a table saw by tilting the blade at 45° and chamfering each corner. Using a lathe, turn the shape on the top of the key block as shown in the *Key Block Detail*.

Note: If you make the key block from a 4 x 4 that's precisely 3½″ square, then four of the faces on the octagon should be 1½″ wide, and the other four 1⁷⁄₁₆″. However, using construction lumber, it's unlikely that you'll find a 4 x 4 that precise. If you can, choose stock that's a little larger than 3½″ and try to make the octagonal faces as close to 1½″ wide as possible.

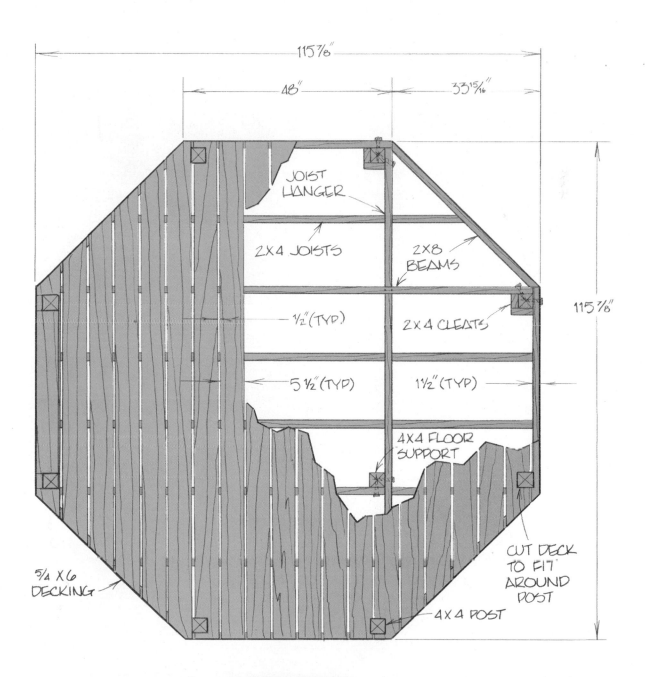

FLOOR LAYOUT

On a flat surface, toenail two rafters to opposite sides of the key block. Temporarily nail a brace between the rafters, a foot or so under the key block. (See Figure 8.) This brace will keep the rafters from pulling out of the block as you raise the roof.

With a helper, put the rafter/key block assembly in place on the gazebo frame, fitting the bird's-mouths over two opposing corners of the cap plate, as shown in the *Roof Frame/Side Elevation.* As the helper holds the assembly in position, toenail the rafters to the cap plate with 16d nails.

Put two more rafters in place, perpendicular to the first pair. Fit the bird's-mouths over the cap plate and lean the mitered ends against the key block. Nail the rafters to both the plate and the block. (The rafters won't support your weight until all eight are in place, so work from a ladder.) At this point, your helper can let go of the rafter assembly — the roof frame should be stable.

Continue to fasten the remaining rafters in place, one at a time. (See Figure 9.) If necessary, chamfer the mitered ends of these last four rafters, as shown in the *Rafter-to-Key-Block Joinery Detail,* so the rafters fit flush against the block. When all the rafters are in place, remove the temporary brace from the first pair.

8/Make a simple truss from two rafters, the key block, and a brace. Nail the brace to the rafter **temporarily** — you'll remove it after you complete the roof frame.

9/As you put each rafter up (after the initial two), nail it to the cap plate **first,** then to the key block. This makes it easier to hold the rafter in place as you work.

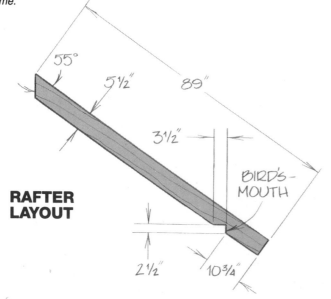

RAFTER LAYOUT

55°
5½"
89"
3½"
BIRD'S-MOUTH
2½"
10¾"

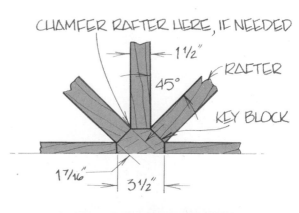

CHAMFER RAFTER HERE, IF NEEDED
1½"
45°
RAFTER
KEY BLOCK
1⁷⁄₁₆"
3½"

RAFTER-TO-KEY BLOCK JOINERY DETAIL

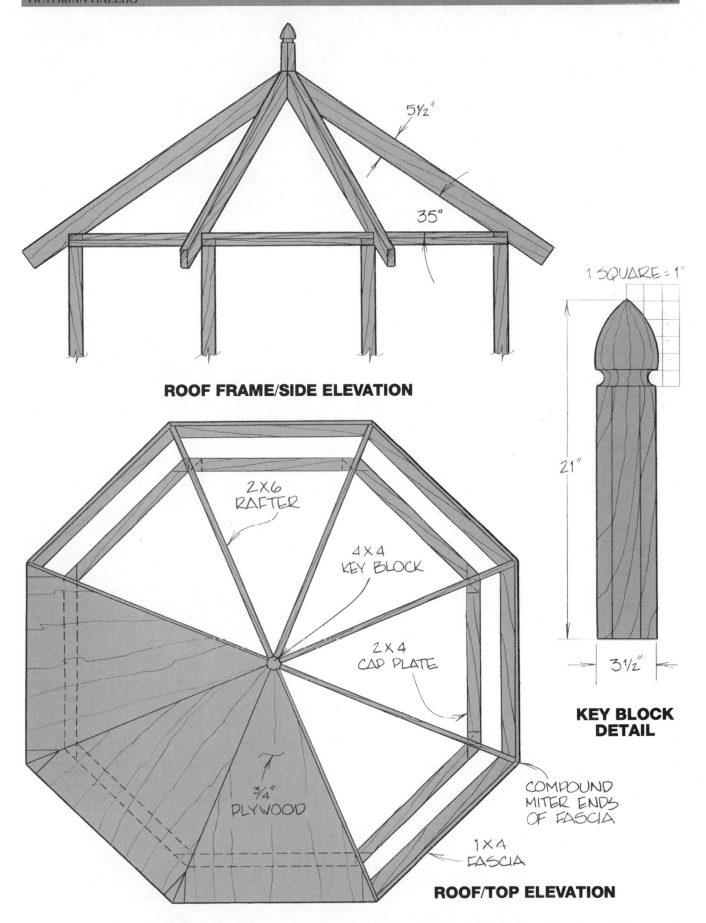

5½"

35°

ROOF FRAME/SIDE ELEVATION

1 SQUARE = 1"

21"

3½"

**KEY BLOCK
DETAIL**

2 X 6
RAFTER

4 X 4
KEY BLOCK

2 X 4
CAP PLATE

¾"
PLYWOOD

COMPOUND
MITER ENDS
OF FASCIA

1 X 4
FASCIA

ROOF/TOP ELEVATION

10 Cut and attach the fascia.

Cut and attach the fascia. Carefully measure each fascia board before you cut it from 1 x 4 stock; the spaces between the ends of the rafters won't be precisely even. Compound-miter both ends of each fascia so it fits tight against the adjoining boards. If you use a radial arm saw to make these miters, tilt the saw blade to 71¾° and swing the arm to 76¾°. For a table saw, tilt the saw blade to 18¼°, and adjust the miter gauge to 76¾°.

Nail the fascia to the ends of the rafters with 8d spiral nails. (See Figure 10.) Don't worry if there are slight gaps between the fascia — it's difficult to get the compound miters to fit precisely on a project this large. Fill any visible gaps with caulk.

*10/Nail the fascia to the rafters **before** you attach the sheathing. This will give the bottom edge of the sheathing something to rest on.*

11 Install the roof sheathing.

Install the roof sheathing. Put up the ¾″ roof sheathing one panel at a time. To measure a panel, put a sheet in place over a section of the roof. Using the *inside* edges of the rafters and the *outside* edge of the fascia to guide your pencil, mark a triangular shape on the plywood. Cut the fascia-side of the triangle precisely along the pencil line, but cut the rafter-sides ¾″ *wide* of the lines. Put the panel back in place on the roof frame and secure it with 8d spiral nails. Repeat for the remaining panels.

A 4′ x 8′ sheet of plywood is not quite big enough to cover an entire panel. So, after installing a sheet, there will be small triangular-shaped openings on either side. After you've attached all eight sheets, cut small pieces of sheathing from the scrap to fill these openings. Nail the pieces to the rafters and fascia with 8d spiral nails. (See Figure 11.)

*11/As you cut each sheet of roof sheathing, save the scrap and **label it** so you know which panel it was cut from — south, southwest, west,* *etc. After cutting and installing all eight sheets, cut triangular-shaped tips off the scraps to cover the small openings on either side of each sheet.* *If you use the scraps from a sheet to cover the openings on either side of that same sheet, each tip will fit precisely.*

12 Install the roofing materials.

Install the roofing materials. Cover the sheathing with roofing paper, using ¾″ roofing nails to hold the paper down. Install drip edge strips around the perimeter of the roof, and cover both the paper and the drip edge with tab shingles. (Refer to "Step-by-Step: Installing Tab Shingles," page 101.)

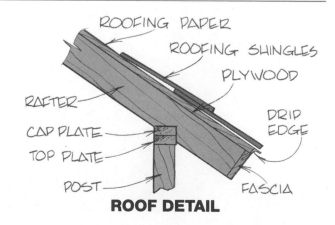

ROOF DETAIL

Install the shingles in rows, parallel to the bottom edge of *each* panel. On the ridges that separate the panels, overlap the shingles. (See Figure 12.) This will look unsightly, but don't worry. Simply install ridge caps to cover the overlapping shingles. (See Figure 13.) Where the shingles meet the key block, fill the seam with roofing compound.

12/As you lay up the rows of shingles, turn the corner at each ridge by overlapping the shingles, as shown.

13/Cover the lapped ends of the shingles with ridge caps. Notice the ladder: When working on the roof, hook the rung of a **metal** ladder over the key block. Use this to keep from slipping off the roof when installing roofing materials. As you finish one area and want to move on to another, pivot the ladder around the block.

13 **Cut and install the railings.** Mount the railings in dadoes cut into the posts. To make these dadoes, first measure the positions of the railings on the posts. The top surface of the top railing should be 33″ above the floor, and the top surface of the bottom railing 5½″. Mark the dadoes on the sides of the posts shown in the *Railing Layout*. Saw a ½″-deep kerf at the top and the bottom of each dado. Remove the waste between the kerfs with a chisel. (See Figure 14.)

Carefully measure the distance between dadoes — it will probably vary slightly from the measurements on the drawings. Cut the railings to fit, mitering the ends of the long railings at 45° as shown in the *Railing Layout*. Place the railings in the dadoes, and secure them to the posts with 10d spiral nails. Drive the nails at an angle through the railings and into the posts.

Remove the points of the mitered railings with a rasp, so the mitered ends are flush with the outsides of the posts. All the railings should be flush with the posts.

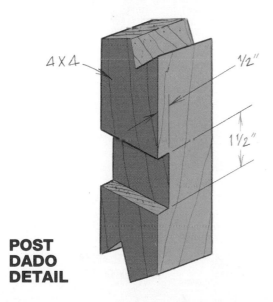

4 X 4

½″

1½″

POST DADO DETAIL

14/To make a dado, set a circular saw to cut exactly ½″ deep, and saw kerfs at the top and bottom of the joint. If the wood is tough, or the joint runs through a knot, cut several more kerfs between the first two. Then remove the waste with a chisel.

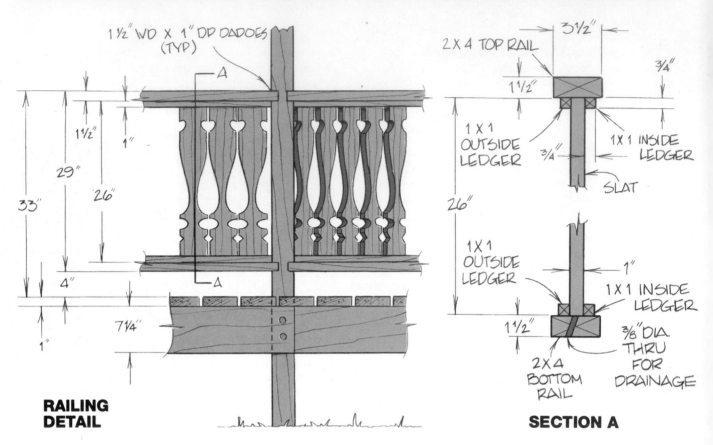

1½" WD X 1" DP DADOES (TYP)

A

1½"

1"

29"

26"

33"

4"

A

7¼"

1"

RAILING DETAIL

2 X 4 TOP RAIL

3½"

1½"

¾"

1 X 1 OUTSIDE LEDGER

1 X 1 INSIDE LEDGER

¾"

SLAT

26"

1 X 1 OUTSIDE LEDGER

1"

1 X 1 INSIDE LEDGER

1½"

2 X 4 BOTTOM RAIL

⅜" DIA. THRU FOR DRAINAGE

SECTION A

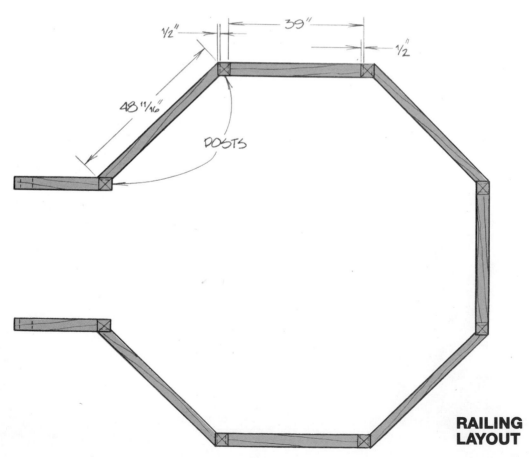

½"

39"

½"

48 ¹¹⁄₁₆"

POSTS

RAILING LAYOUT

14 Cut the slats and install them between the rails.

From 5/4 x 6 decking stock, rip and cut 60 slats, 4¼″ wide and 26″ long. Enlarge the *Slat Pattern* and trace it on a scrap of hardboard. Cut the hardboard on a band saw to make a template, then use this template to mark the pattern on 15 to 20 slats.

Stack 3 or 4 slats, with one of the marked slats on top. Tape the stack together, then cut the pattern on a band saw. When you remove the tape, you'll have several identical slats. Repeat, cutting the same pattern in all the slats.

Rip ¾″ x 1″ ledgers from 1 x 4 stock. Cut the outside ledgers to fit between the posts, underneath the top rail or on top of the bottom rail, as shown in *Section A*. Miter the ends or cut them square, as needed, so they rest flush against the posts. Nail the outside ledgers in place with 6d common nails.

Place the slats between the top and bottom rails, spaced ½″–¾″ apart and against the outside ledgers. There should be room for eight slats between each set of square-cut rails, and nine between each set of mitered rails. Adjust the positions of the slats until they appear to be spaced evenly. Then toenail them in place, driving 6d common nails at an angle through the slats and into the rails.

Cut inside ledgers to fit between the posts, against the slats. Once again, miter the ends or cut them square, as needed. Put the ledgers in place against the slats and attach them to the rails with 6d common nails. (See Figure 15.)

> **TRY THIS!** It's hard to paint the slats once you've installed them, because they're so close together. You must use a small brush to reach the edges, and this takes lots of time. The job will go more easily if you paint the slats *before* you put them in place.

The inside and outside ledgers on the bottom rails create a gutter that will fill up with rainwater. To prevent this, drill ⅜″-diameter drainage holes down through the bottom rails wherever there's a gap between two slats or between a slat and a post. To reach between the slats with the drill, you'll have to make these holes at an angle, as shown in *Section A*.

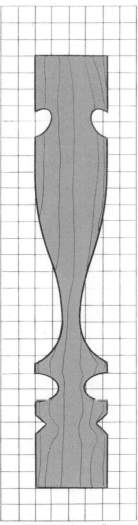

SLAT PATTERN

1 SQUARE = 1″

15/The slats are kept in place by ledger strips, top and bottom, inside and outside. Nail the outside ledgers in place first and place the slats against them. Then attach the inside ledgers.

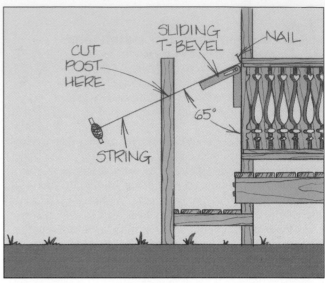

15

Attach the stair railings. The stair railings should be parallel to the stairs' *angle of descent* — approximately 65° on this structure. Use this angle to determine where to cut off the newel posts. Drive a small nail into one of the supporting posts above the stairs, even with the *bottom edge* of the top rail. Tie a string to the nail. Stretch the string taut between the posts, touching the inside faces. Have a helper move the free (lower) end of the string up and down while you measure the angle between the string and the supporting post. When that angle is equal to 65°, have the helper mark the position of the string on the newel post. (See Figure 16.)

Repeat this process for the other newel post. With a string and a string level, check that your marks on both newel posts are even with each other. If not, repeat your measurements and move one or both marks as necessary. Cut off the posts at the marks, following the 65° layout lines.

Cut the stair railings, mitering both ends at 65°. Put the railings in place, checking that the lower end of each railing overhangs the newel post by approximately 2″, as shown in the *Stair Frame/Side Elevation*. Attach the stair railings to the posts with 10d spiral nails.

16/Use a string and a sliding T-bevel to determine where to cut the newel posts. Set the T-bevel to 65° with a protractor. Stretch the string *between a supporting post and a newel post, as shown, and measure the angle between the string and the supporting post with the* *T-bevel. Move the string up and down on the newel post until the angle of the string matches the angle of the T-bevel.*

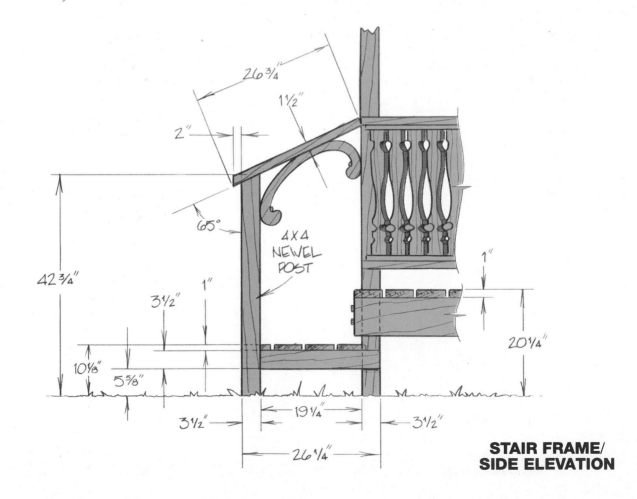

STAIR FRAME/ SIDE ELEVATION

16 Cut and attach the ornaments.

Enlarge the *Corner Ornament Pattern, Key Block Ornament Pattern,* and *Heart Ornament Pattern* and make three hardboard templates. Mark the shape of the ornaments on 2 x 10 stock, and cut them on a band saw. You'll need 16 corner, 4 key block, and 4 heart ornaments.

You must bevel the corner ornaments on the edge that meets the posts. To cut this bevel, save one of the scraps you cut from *inside* the curve of a corner ornament. Attach this scrap to the miter gauge of a table saw, and use it to hold the corner ornaments while you cut the bevels. (See Figure 17.)

Fasten the ornaments in place with lag screws, as shown in the *Ornament Joinery Details.* Use ⅜" x 4" lag screws to attach the corner ornaments to the posts and top plates (see Figure 18) and the key block ornaments to the key block (see Figure 19). Drive the screws through the ornaments and into the gazebo frame. Use ⅜" x 6" lag screws to attach the hearts, driving the screws through the 2 x 4 plates and into the ornaments.

TRY THIS! Like the slats, the ornaments will be difficult to paint once you attach them to the frame. You can save a lot of work by painting the ornaments *before* you install them.

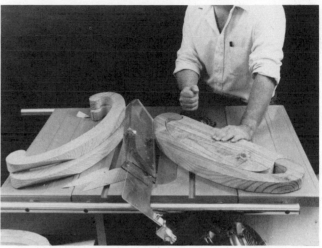

17/Save the scrap from a corner ornament to hold these ornaments when beveling the edges. Angle the miter gauge to 45° **and** tilt the blade to 45°. Place the scrap against the face of the gauge and fit the ornament over the scrap. Slide the scrap back and forth on the gauge until the ornament is positioned properly on the saw, then fasten the scrap to the miter gauge. Cut the bevels on **four** ornaments. Then turn the miter gauge around (front for back) in the table slot, reposition the scrap on the gauge, and cut the remaining four bevels.

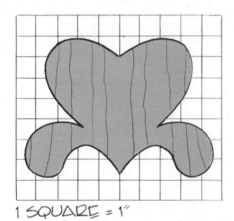

1 SQUARE = 1"

HEART ORNAMENT PATTERN

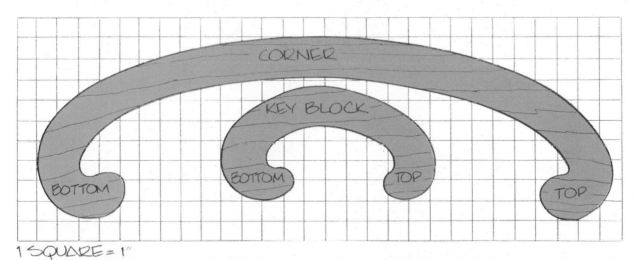

1 SQUARE = 1"

CORNER ORNAMENT PATTERN AND KEY BLOCK ORNAMENT PATTERN

18/To attach a corner ornament, hold it in place and drill a 5/16"-diameter pilot hole through it and into the gazebo. Put a flat washer over the end of a 3/8" x 4" lag screw, and drive it through the ornament and into the post or frame. Fasten each corner ornament with two lag screws.

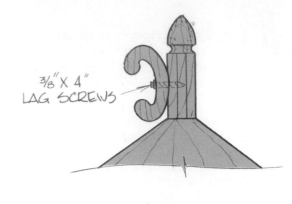

3/8" X 4" LAG SCREWS

3/8" X 6" LAG SCREWS

3/8" X 4" LAG SCREWS

MITER EDGE AT 45°

ORNAMENT JOINERY DETAILS

19/When you attach the key block ornaments, you won't be able to use the ladder to keep your footing on the roof. Instead, nail scraps of 2 x 4 to the roof panels and brace your feet against these. When you've attached all four ornaments, remove the 2 x 4s. On the first hot day, the asphalt shingles will soften and the nail holes will close.

3/8" X 4" LAG SCREWS

17 Paint the gazebo. Let the structure sit for several weeks, through at least one drenching rain, to wash away any chemical residue on the treated lumber. The wood also needs some time to season. All the lumber — both treated and untreated — will be slightly green when you buy it. Even kiln-dried construction lumber can have as much as 25% to 30% moisture content. Let some of this moisture evaporate before you paint the wood.

When you judge the structure has set long enough, prime the raw surfaces of the wood. Then paint the gazebo, inside and out. Traditionally, Victorian garden structures were painted white or a light pastel color. (The gazebo shown is white.) Paint yours any color that you fancy.

Credits

Contributing Craftsmen and Craftswomen:

Phil Baird (Sunshade)

Dan Callahan (Sheltered Firewood Rack)

Larry Callahan (Country Deck, Fences and Gates)

Nick Engler (Multipurpose Frame Building, User-Friendly Doghouse, Custom-Built Birdhouses, Fences and Gates, Trash Can Corral)

Mary Jane Favorite (Victorian Gazebo)

Special Thanks To:

Rodney Barret
John Stephens
Robert Walendzak
Tony Walendzak
Patterson's Flowers, West Milton, Ohio
Wertz Hardware Stores, West Milton, Ohio

Chairs and glassware for cover photograph supplied courtesy of Provincial House, 2159 Valley Street, Dayton, Ohio.

Rodale Press, Inc., publishes AMERICAN WOODWORKER™, the magazine for the
serious woodworking hobbyist. For information on how to order your subscription,
write to AMERICAN WOODWORKER™, Emmaus, PA 18098.

WOODWORKING GLOSSARY

Parts of a Board

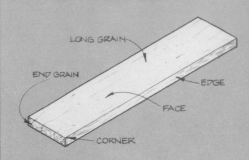

LONG GRAIN
END GRAIN
EDGE
FACE
CORNER

Basic Saw Cuts

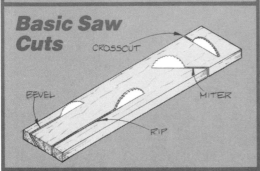

CROSSCUT
BEVEL
MITER
RIP

Parts of a Drawer

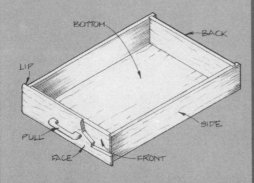

BOTTOM
BACK
LIP
SIDE
PULL
FACE
FRONT

Parts of a Frame

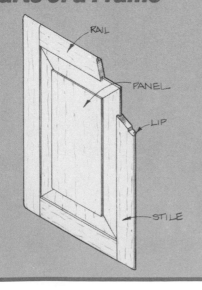

RAIL
PANEL
LIP
STILE

Basic Joinery

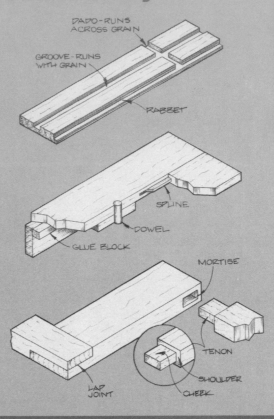

DADO-RUNS ACROSS GRAIN
GROOVE-RUNS WITH GRAIN
RABBET
SPLINE
DOWEL
GLUE BLOCK
MORTISE
TENON
SHOULDER
LAP JOINT
CHEEK

Parts of a Tab[le]

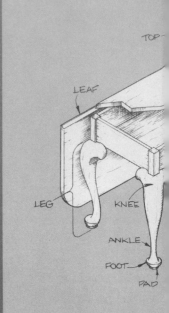

TOP
LEAF
LEG
KNEE
ANKLE
FOOT
PAD

Common Shapes and Moldings

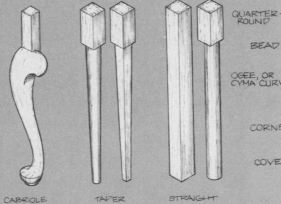

QUARTER-ROUND
BEAD
OGEE, OR CYMA CURVE
CORNER
COVE
BED
CROWN

CABRIOLE
TAPER
STRAIGHT

Pa[rts]

FACE FRAM[E]
WEB FRAM[E]
SHEL[F] SUPP[ORT]
BASE

Holes

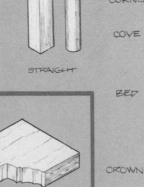

SCREW HOLE
STOPPED HOLE
THRU HOLE
COUNTERBORE
COUNTERSINK
PILOT HOLE